T0224986

SpringerBriefs in Applied Sciences and Technology

Forensic and Medical Bioinformatics

Series Editors

Amit Kumar, Dwarka Venkat Sai Nagar Colony, Munaganoor, Hayatnagar, BioAxis DNA Research Centre Private Ltd, Hyderabad, Telangana, India

Allam Appa Rao, Hyderabad, India

More information about this subseries at http://www.springer.com/series/11910

Kacper Choromanski

Bloodstain Pattern Analysis in Crime Scenarios

 Springer

Kacper Choromanski
University of Warsaw
Warsaw, Poland

ISSN 2191-530X ISSN 2191-5318 (electronic)
SpringerBriefs in Applied Sciences and Technology
ISSN 2196-8845 ISSN 2196-8853 (electronic)
SpringerBriefs in Forensic and Medical Bioinformatics
ISBN 978-981-33-4427-3 ISBN 978-981-33-4428-0 (eBook)
https://doi.org/10.1007/978-981-33-4428-0

This Springer imprint is published by the registered company Springer Nature Singapore Pte Ltd.
The registered company address is: 152 Beach Road, #21-01/04 Gateway East, Singapore 189721, Singapore

Preface

Reconstruction of the criminal events based on evidence is a subject that appears on TV and in books very often. How often you can see genius detective who will read crime scene like a book and will give detailed presentation what really happen? But how does it look like in the real world. How experts from this field reach their conclusions? What techniques or terminology they are using? And what is the most important: is this method useful or is it just a "junk science"? What kind of mistakes experts can made? This book will try to answer those and other question.

Warsaw, Poland   Kacper Choromanski

Acknowledgements

First of all, I want to thank my great friend and mentor Stuart H. James. He was the person who introduced me to the BPA, and I will be thankful for that for the rest of my life. Thank you Stuart! Furthermore I want to acknowledge the support and encouragement that we received throughout my years of research on bloodstain pattern analysis from University of Warsaw (Faculty of Law & Administration and the Centre for Forensic Sciences), prof dr hab. Piotr Girdwoyń, prof dr hab. Tadeusz Tomaszewski, dr hab. Kacper Gradoń.

Also, I want to thank National Science Centre (NCN), Poland for being a Prime Investigator for "PRELUDIUM 16" research project, ref. no UMO-2018/31/N/HS5/01031. Without this support, one of the chapters would be much shorter and shallow.

I want to give huge thanks to my friend Rafał Starosielec for support, numerous comments and a great help as an english corrector with this book.

Furthermore, I want to thank dr hab. Andrzej Witowski, dr Michał Dobrowolski, Piotr Słowinski. Your numerous kind words about my research and almost unlimited time for scientific discussions are precious.

Last but not least I'm sending my full regards to my Family, daughter Debra, son Hugo and wife Kasia for their patience and understanding. Without you, this book would not be possible. Thank you for being such a great inspiration for improving myself. I love you all.

Contents

Chapter 1
Introduction

Abstract Introduction to role of forensic science in criminal cases.

Keywords Crime · Forensic · Casework · Bloodstain pattern analysis

Crime and violence were is and will be present in human history. Peace or war, outside and inside of the borders, among enemies and families, crime is a current problem. Thankfully technique, research, science is driven by human goodwill is a significant force, that tries to handle with this phenomena. Armaments race between criminals and law enforcement authorities is still an open case. Criminals have to come up with new ideas how to commit a crime, and the police with forensic experts have to figure out new techniques and methods to search, locate, identify and explain how particular criminal even was done. More advanced science also leads to more questions about practices themselves. Nowadays, problems with miscarried justice, over the interpretation of the evidence, and junk science is more visible than ever. The broad spectrum of techniques that can be used during investigations gives investigators great opportunities to understand events better and explain them in more details than before. Fingerprint identification, footprint identification, anthropology, forensic medicine, forensic ballistics, toxicology or forensic DNA identification, named few of forensics sciences that give a crucial input to the real casework. There are also forensic sciences that are focusing on the reconstruction of the event like ballistics, reconstruction of the car crashes or bloodstain pattern analysis.

In this book, I would like to explain how bloodstain pattern analysis work, both in theory and in practice. Several excellent books have been written about bloodstain pattern analysis. I just would like to give my contribution to the forensic community by this work by showing how science can be beneficial for crime-solving scenarios. This book will focus on that matter. I hope that examples that I will show on the next pages will help to understand the basics of bloodstain pattern analysis and its role in today's forensic practice.

On of the high-point of this book is an examination of real case scenarios, homicide and suicide case. Both investigations are well documented not only by description and reconstruction but also by real crime scene photography.

Information that I will be sharing in next pages, will be helpful not only for police or expert. This book is an excellent source of knowledge for the practitioners, prosecutors, attorneys, judges and everyone interested in forensic science and case-solving.

Chapter 2
Basics of Bloodstain Pattern Analysis

Abstract Bloodstain pattern analysis is a field of forensics, that allows you to determine the courses of events. Basic terminology, issues related to these patterns and the methodology of such forensic investigation will be laid down in the following chapter.

Keywords Bloodstain pattern analysis · Practice · Reconstruction of the event · Terminology · Methodology · Practice

Blood is a biological evidence which is most commonly found and secured on the crime scenes (James et al. 2005). It is a common knowledge that genetic testing of such traces allows to pinpoint a person from which the tissue came. Unfortunately, fewer people, including forensic investigators, are aware of the fact that you can also examine these patterns in order to determine how they were created. Bloodstain pattern analysis (BPA) is concerned mostly with this issue. This book aims to give you a more concrete knowledge on the matter, in (hopefully) an interesting and easily accessible way.

So what is Bloodstain Pattern Analysis? It's a field of forensic science concerned with examining bloodstain patterns. You can determine the way in which the bloodstain patterns were created by analyzing their sizes, shapes, localization and placement in relation to other stains. Having this information, you can determine the course of events, that led to the creation of these stains. Such information can be crucial for further work of the investigators. It can help to either confirm or rule out certain scenarios or verify the version of events presented by a witness, a suspect or a victim.

Contrary to popular belief, one of the most beneficial outcomes of a properly conducted bloodstain pattern analysis is the ability to rule out certain possible scenarios presented to the investigators. This may seem counterintuitive, but it brings measurable benefits in the long run.

First and foremost, a properly conducted bloodstain pattern analysis may rule out someone's participation in the crime, which can save a truly innocent person from the criminal liability for the crimes he or she didn't commit. It is a huge benefit, especially in the face of recently popular, and often publicized, cases of unjust conviction.

Second of all, even when the law enforcement authorities have a suspect, a properly conducted bloodstain pattern analysis may help, by ruling out certain scenarios— stories presented by a suspect that contradict with the results of the conducted analysis. The realization of investigators' "omniscience" regarding the events, which led to exsanguination, may make enough of an impression on the suspect, that he will more willingly cooperate with law enforcement and judicial authorities.

Thirdly, during cases in which the investigators are still looking for a suspect, they need to verify various scenarios. If certain versions of events can be ruled out as a result of forensic science, the investigator's work will be that much easier. From the second in which they receive such information, they can focus strictly on the realistic scenarios, instead of wasting time on the versions of events leading nowhere. This way, bloodstain pattern analysis may measurably benefit the efficient utilization of financial resources, time resources and manpower.

If the investigators are able to correctly determine both the sequence and the course of events, they could more precisely establish the legal qualification of the conduct. A prosecutor in charge of such a case will differently formulate an indictment, for what he thought was a scuffle and a brawl, and differently if he had more detailed data. Let's assume a scenario in which, as a result of BPA investigation, it was possible to determine that not only was there a scuffle and a brawl but also that the suspect was repeatedly kicking the victim's head as he was lying on the ground. This seemingly insignificant detail may determine whether the act would be treated severely or not. In this scenario, instead of a brawl, the assailant may be indicted for attempted murder. Naturally, such legal details depend on the penal system in which the cases are processed, but it's important to highlight that bloodstain pattern analysis may provide a wide range of possibilities in that field.

A possibility of determining a course of events due to BPA is not only applicable in homicide cases. Upon discovery of a dead body, the investigators may also encounter a case of suicide, a death by natural causes or an accident. It's also important to highlight that not in all cases of exsanguination there has to be a corpse. Injuries, nonfatal accidents, failed suicide attempts, batteries, kidnappings, tortures, abuses and other similar cases are also a possibility. Depending on the secured evidence, the experts may be able to determine, to some degree, the course of events, which may prove helpful either to the work of either the investigators or the judicial authorities. Also, it must be noted that these pieces of information may also be relevant in civil lawsuits, insurance issues and other cases in which, for example, a description of an accident may have a huge significance for the case.

2.1 Interdisciplinarity

Bloodstain pattern analysis is rooted in a few different fields of science (Bevel and Gardner 2008; Choromanski 2020). It has to be kept in mind especially when you're the expert in these fields, drafting an opinion. That fact can also prove to be important if you're a prosecutor or a judge, since when you appoint someone as an expert for a specific case, you have to be aware of the complex nature of the opinion given, as well as a broad scope of information presented in it. Bloodstain pattern analysis, a forensic science determining the way in which bloodstain patterns are formed, has its roots in the field of mathematics, physics, chemistry and biology. Analysis has to be carried out within a legal framework, which is a legal codification of a country or a region in which it takes place, which is why it should be noted that the field of law is also relevant for the studies.

What exactly is the influence of different fields of science on the bloodstain pattern analysis?

Physics helps to understand the formation process of a droplet, its movement in the air, the movement of blood volumes in space. Thanks to physics, it is possible to determine possible changes in the position of objects in the event based on flow pattern localised on those objects. The shape of this pattern can determine the direction of the gravitation force. Additionally, thanks to the use of various types of light (Sterzik et al. 2016), using the benefits of physics, it is possible to reveal previously hard to notice bloodstains.

In addition, knowledge of this field of physics allows for a better understanding of the process of blood drying.

Mathematics will enable analysts to calculate the angle of impact of a drop on a given surface (Larkin and Banks 2013). The geometry will allow you to determine the approximate flight path and thus determine the place where was the area of origin of the spatter (Conlan et al. 2019). Also, mathematics enables the use of procedural algorithms and help with a better understanding of probabilistic calculus (Matisoff and Barksdale 2012; Camana 2013).

Biology is fundamental in bloodstain pattern analysis. First and foremost, biological sciences allow identification of whether a given substance is blood or whether it is just a russet-red substance. In addition, it enables you to understand the processes taking place with the blood inside the body and the processes that occur when blood is outside the bloodstream. The analysis of diseases and factors that have a direct impact on the physicochemical properties of blood also plays an immense role. Also, the use of genetic sciences in conjunction with inference about bloodstain pattern analysis allows for categorical interpretation and the creation of detailed reconstructions of the event.

Chemistry is no less important than all the disciplines mentioned so far. Thanks to the use of chemical processes, investigators can reveal and document blood stains that are invisible at first glance (Praska and Langenburg 2013; Hong and Seo 2015; Al-Eid et al. 2018; Chingthongkham et al. 2020). The main principle of applying chemistry at the scene is the use of the contrast. A red stain on a red background will be difficult to see, but if the stain turns black under the influence of a chemical, the colour difference will be a considerable giving visible stain. Another function of using chemistry in bloodstain pattern analysis is to perform preliminary tests to see if the trace can be blood or not. It is worth mentioning that after the preliminary tests, more detailed biological tests should be performed categorically stating that the test stain is human blood.

2.2   The Consequences of Different Stain Localization

The fact that the bloodstain pattern was revealed somewhere or on someone doesn't imply a direct connection with a committed crime. You have to keep in mind that blood is subject to two form shifting processes. Coagulation and drying up. Drying up is a physical process consisting of the evaporation of water from the stain. The coagulation process, however, is a biochemical process, resulting in changing from a liquid to a gel. These two processes are interconnected with each other and can occur simultaneously, resulting in dried up blood clots. There is no universal measurement of time indicating when exactly blood will change from liquid form to solid. Considering the aforementioned facts, you have to carefully approach the matter of precisely estimating the time, in which the given bloodstain was created. It also must be noted that blood in its liquid form may create other patterns when it comes in contact with other surfaces.

Bloodstain patterns can be found in a variety of places, such as:

– place where the crime was committed
– place where the body was found
– place where participants of the events resided or passed through
– objects used for transport by the participants of the events, such as:

  • cars
  • motorcycles
  • vans
  • lorries
  • boats
  • motorboats
  • trunks
  • trailers
  • and other similar objects

– directly on the participants of the events
– on objects used by the participants of the events
– on clothes worn by the participants of the events

It must be noted that participants of the events include the victim, the assailant and the witnesses. During the events, there is a possibility of transferring bloodstain patterns onto a third party, which was unrelated to the criminal incident, for example:

– paramedics making contact with the victim's blood, while giving first aid
– family member making contact with blood on a doorknob, while discovering the corpse of his kin
– forensic investigators making contact with blood, while performing the crime scene investigation
– unaware janitor dirtying a mop, while using a bucket which previously contained blood

- someone's contact with blood while evaluating the medical condition of a person laying on the ground
- firefighters making contact with the blood of a car crash victim, while attempting to open said vehicle

Naturally, during such actions blood may be transferred not only onto other people, but also objects and the scene itself. As a result of interacting with the victim's body, it is possible to create new bloodstain patterns on the victim, which will have serious consequences during the subsequent bloodstain pattern analysis. The very fact of revealing bloodstain patterns on a person or an object doesn't necessarily imply that you're dealing with the assailant or the murder weapon. The aforementioned examples unequivocally show that blood can be relatively easily transferred, even onto the innocent; those that have never been involved in the crime.

The abovementioned argument proves that bloodstains secured during the forensic examination should be investigated in two ways. Firstly, identified biologically and genetically—thanks to the results of these tests, you can categorically state, whether a given bloodstain pattern is human blood, and if so, then to whom does it belong. The second type of test required in order to get to the truth of the events is analyzing how the bloodstains were created in the first place, which is the domain of bloodstain pattern analysis. Only a bloodstain analyzed in these two ways will be of use to the investigators. A comprehensive and meticulous investigation may yield a satisfying result and determine what really had happened.

2.3 Terminology

Uniformity of terms, categorization and processes is a key element of any science (Choromański 2017). Forensics, including bloodstain pattern analysis, is no different in this regard. Some pieces of literature highlight the importance and necessity of uniform terminology (Choromański 2013). There are numerous advantages that come with its implementation. First of all, uniform terminology allows for a better understanding of the subject matter. Science becomes consistent if a given word or phrase precisely describes any phenomenon, e.g.: a way in which a stain was created, or a characteristic trait of a bloodstain pattern. In such a situation, two different people will be able to fluently communicate, while looking at the same stains, limiting a possibility of errors or miscommunications. Not only forensic experts would benefit from this, as such change would also impact prosecutors, defenders and judges. Uniform terminology also allows for an easier verification of expert's opinions. A properly defined vocabulary would make understanding of the subject matter much easier. And speaking of subject matter, another relevant element of uniform terminology springs to mind. Creation of a universal glossary gives an international capacity to any given field of forensics, which, in turn, facilitates international cooperation and research of subject matter on basic, developmental and applied levels. It creates a limitless communication platform, resulting in two experts, on the opposite sides of the globe, reaching the same conclusion while looking at a single piece of evidence. Uniform terminology should be a standard, and fortunately, there is a glossary for BPA. The first attempt of such unification was made by a group of experts from North America, Europe and New Zealand called SWGSTAIN (Scientific Working Group on Bloodstain Pattern), which was founded in March 2002 at a meeting held by the FBI Laboratory at the FBI Academy in Quantico, Virginia. The result of their work was a glossary of basic, recommended terms, published in 2009 (Forensic Sci. Commun. 11 (2009)). They also worked on development of bloodstain pattern analysis, its terminology, evaluation of applied methods, techniques, rapports and quality control in this particular field of forensics. Later on, a group called OSAC (Organization of Scientific Area Committees) took over the work on the terminology, improving and correcting SWGSTAIN's efforts. In following pages, I will present the terminology recommended by OSAC, which currently is the newest recommended glossary in BPA field.

2.4   OSAC Terminology of Bloodstain Pattern Analysis

I.1 Accompanying drop- A small blood drop produced as a by-product of drop formation.

I.2 Altered stain- A bloodstainwith characteristics that indicate a physical change has occurred.

I.3 Angle of impact- The angle (alpha), relative to the plane of a target, at which a blood drop strikes the target.

I.4 Area of convergence—The space in two dimensions to which the directionalities of spatter stains can be retraced to determine the location of the spatter producing event.

I.5 Area of origin- The space in three dimensions to which the trajectories of spatter can be utilized to determine the location of the spatter producing event.

I.6 Backspatter pattern- A bloodstain pattern resulting from blood drops which can be produced when a projectile creates an entrance wound.

I.7 Blood clot- A gelatinous mass formed by a complex mechanism involving red blood cells, fibrinogen, platelets, and other clotting factors.

I.8 Bloodstain- A deposit of blood on a surface.

I.9 Bloodstain pattern- A grouping or distribution of bloodstains that indicates through regular or repetitive form, order, or arrangement the manner in which the pattern was deposited.

I.10 Bubble ring An outline within a bloodstainresulting from air in the blood.

I.11 Cast-off pattern-A bloodstain pattern resulting from blood drops released from an object due to its motion.

I.12 Cessation pattern- A bloodstain pattern resulting from blood drops released from an object due to its abrupt deceleration.

I.13 Directionality- The characteristic of a bloodstainthat indicates the direction blood was moving at the time of deposition.

I.14 Directional angle- The angle (gamma) between the long axis of a spatter stainand a defined reference line on the target.

I.15 Drip pattern- A bloodstain pattern resulting from a liquid that dripped into another liquid, at least one of which was blood.

I.16 Drip stain- A bloodstain resulting from a falling drop that formed due to gravity.

I.17 Drip trail- A bloodstain pattern resulting from the movement of a source of drip stainsbetween two points.

I.18 Edge characteristic- A physical feature of the periphery of a bloodstain.

I.19 Expiration pattern- A bloodstain patternresulting from blood forced by airflow out of the nose, mouth, or a wound.

I.20 Flow- A bloodstain resulting from the movement of a volume of blood on a surface due to gravity or movement of the target.

I.21 Forward spatter pattern- A bloodstain pattern resulting from blood drops which can be produced when a projectile creates an exit wound.

I.22 Impact pattern- A bloodstain pattern resulting from an object striking liquid blood.

I.23 Insect stain- A bloodstain resulting from insect activity.

I.24 Parent stain- A bloodstain from which a satellite stain(s)originated.

I.25 Perimeter stain- An altered stain consisting of its edge characteristics, the central area having been partially or entirely removed.

I.26 Pool- A bloodstain resulting from an accumulation of liquid blood on a surface.

I.27 Projected pattern- A bloodstain pattern resulting from the ejection of blood under hydraulic pressure, typically from a breach in the circulatory system.

I.28 Satellite stain- A smaller bloodstain that originated during the formation of the parent stainas a result of blood impacting a surface.

I.29 Saturation stain- A bloodstain resulting from the accumulation of liquid blood in an absorbent material.

I.30 Serum stain- The stain resulting from the liquid portion of blood (serum) that separates during coagulation.

I.31 Spatter stain- A bloodstain resulting from an airborne blood drop created when external force is applied to liquid blood.

I.32 Splash pattern- A bloodstain pattern created from a large volume of liquid blood falling onto a surface.

I.33 Swipe- A bloodstain resulting from the transfer of blood from a blood-bearing surface onto another surface, with characteristics that indicate relative motion between the two surfaces.

I.34 Target- A surface onto which blood has been deposited.

I.35 Transfer stain- A bloodstain resulting from contact between a blood-bearing surface and another surface.

I.36 Void- An absence of blood in an otherwise continuous bloodstain or bloodstain pattern.

I.37 Wipe- An altered stain resulting from an object moving through a preexisting wet bloodstain.

2.5 Methodology

Scientific methodology, upon which bloodstain pattern analysis is based, can be divided into two categories. The first category include developing and conducting experiments, in order to recreate the bloodstain patterns encountered on crime scenes, objects or people. The second category focuses on drawing conclusions based on bloodstain patterns in real cases, by comparing found bloodstain patterns with those created through aforementioned experiments, in order to recreate the mechanisms creating them, resulting in a meaningful reconstruction of the events. Both categories are equally important. As it was previously mentioned, the first methodology concerns experiments related to bloodstain pattern analysis. The results of said experiments are a basis for further concluding. They have their roots in very basic mathematical, physical, chemical and biological theories. The main areas, around which they revolve, are Newtonian physics, gravitational force, centrifugal force and kinetic energy. There are more advanced problems concerning fluid dynamics or mechanics of clot formation, alongside factors impacting them. However, they present a sort of boundary of information, which is constantly pushed further through new research, which will be discussed in a different chapter. The most basic experiments, conducted by every trainee in the field of bloodstain pattern analysis, concern the distinctions between specific bloodstain categories/types. During the first few training sessions, the correlation between the angle of drops trajectory and the stain's shape is explored. Furthermore, the trainees reconstructing all possible stain shapes. These experiments are conducted in nearly sterile conditions, in comparison to the places in which the experts work. The substance used is either human, animal or artificial blood. In the first two cases, blood is mixed with anticoagulants, in order to prevent clotting.

The second methodology concerns the very process of drawing conclusions. The proper steps should be done during bloodstain pattern analysis (Gardner 2006). First and foremost, I need to stress one aspect, which is often omitted in scientific literature. Conclusions can be drawn only on stains that were verified as bloodstains through biological or genetic tests. Otherwise, the investigators deal with brown and red substance, with physicochemical properties similar to blood's. This is absolutely crucial for proper inference. Unfortunately, as I've frequently seen in practice, when discussing a stain, experts tend to state that it is in fact human blood, with no biological tests to prove it, in order to draw far-reaching conclusions. There is no scientific basis for such reasoning, as it is solely based in a misleading assumption. It is a very basic mistake, which I've encountered on numerous occasions in my career. It's worth mention that so far there isn't established one proper methodology of giving detailed conclusions. Every expert has to explain how he works. For my best knowledge at this moment, there is an enormous effort in the bloodstain pattern analysis community to establish proper methodology for giving proper conclusions.

In case of stain categorization, and the following reconstruction of events, the forensic methodology is in a direct opposition to the so-called Ockham's Razor. Ockham's razor principle is to limit the creation of the so-called entities. After all, the main task of science is to explain the phenomenon in the most accessible and easy way. Why, in my opinion, this principle of Ockham's razor does not support bloodstain pattern analysis? Well, it seems to me that science should actually be categorical on issues. But sometimes one answer is not enough (Latham 2011). Sometimes one phenomenon can be explained by two or more ways. In an ideal world, Ockham's razor should give only one, single most probable answer. Unfortunately, this is not the case with the bloodstain pattern analysis. At the moment, not all the answers are known, and this field of forensics raises further surprising questions. I will write more on this in the last chapter of this book, but I would like to emphasize it here. If there is a stain which with its appearance, shape, location and arrangement indicates that it could have arisen in at least two similar but still different mechanisms, an expert should write about this. Is it right to narrow the answer down to one most likely mechanism when others cannot be excluded? In my opinion, no. Of course, this favours Occam's razor principle, because we limit ourselves to a single specific answer. However, in such a scenario, it does not implement the basic assumption of science, which is to answer the question comprehensively. Such a case can be said to be half true. Reconstructing an event involves creating possible scenarios or excluding them, and this is what you should remember.

2.6 Documentation of Bloodstains

Creating detailed documentation of bloodstain pattern at the crime scene is one of the most critical tasks required to solve most cases. Documentation of the appearance, location, size and alignment of objects and traces helps to understand the space and evidence better. Without this information, it is impossible to recreate the event that resulted in the creation of documented evidence.

Of course, the method of documentation and its requirements will be various in different jurisdictions. The means and tools by which this documentation is implemented is a broad group of activities. These include written forms in the form of a detailed description, sketch and technical dumps. There is forensic photography (Gouse et al. 2018), spherical photography and 3D scanners. As an analyst of bloodstains, I appreciate all forms, but if I were to choose the most effective and economically accessible to most individuals, for today, I would select forensic photography. However, it is worth noting that the forms listed above complement each other. The photo and the scanner cannot include some of the information that can be recorded in the written report and vice versa. The picture does not capture the temperature, humidity, condition whether the trace is dry or wet, the presence of air draft or other factors. Therefore, the golden rule should be to use a hybrid form in order to secure information about the traces investigated by investigators comprehensively. The hybrid form is based on the accompanying documentation of traces in various ways, the most popular being photography combined with written reporting. Coming back to 3D scanners, at the moment they are not used in a wide range, but they are becoming more and more popular. Their price is declining, making them economically affordable for an even more comprehensive range of experts, units and police groups.

As mentioned earlier, forensic photography is now the most popular form of documenting bloodstain patterns. Correctly carried out the process of securing the appearance of the evidence focuses on a few simple rules. Importantly, these rules apply to the methodology of taking photos and the appropriate algorithm of the procedure. They do not include the hardware itself. This means that an expert with a straightforward tool can take valuable images. It may also happen that with great equipment, and we will not apply the rules that I will write about in a moment. This will result in little valuable evidence, which may make the reconstruction of the event limited or even impossible.

2.7   Principles of Photographic Documentation of Bloodstain

In order to adequately preserve and document the bloodstain pattern, investigator should follow the standards of forensic photography. These are just a few rules, but their application significantly improves the quality of the evidence. For now, I will be on taking a single photo of a single spot/bloodstain. Then I will discuss how to take a series of images, several patterns located in different places.

In the case of single evidence, make sure that the surface with the visible stain is at an angle of 90 degrees to the camera lens. As a result, the created image should be relatively least distorted by the perspective.

Then, in addition to the pattern, the photo should contain a scale, thanks to which anyone who will see the picture will know what the real size of the evidence is (Ferrucci 2016).

Another rule of thumb for taking pictures of single evidence is that the stain or pattern should account for most of the surface of the photograph. It can be achieved with conventional zoom.

When photographing a single spatter or other type of stain, there should be added to surface the so-called baseline. It is the line parallel do the ground level. Thanks to this information, it is known at what angle the area of origin of the spatter is located in relation to the ground-level/baseline. Based on that it's easy to establish ether it is directed downwards or upwards. The baseline is a line which is also perpendicular to the direction of gravity. This information is necessary because when you zoom in, for single spatter stain, the perspective will lose lots of other information. Focusing on single spatter stain without addition marks will lead to lesser details about the position of this trace in relation to the level.

The last rule in the case of photographs of a single pattern is that the photo should contain the order number of given evidence, which is consistent with what is described in the report/sketch/technical projection.

In the case of photographs of more massive clusters of stains, remember to first photograph from a broad perspective. Thanks to this, it will be possible to include as many traces as possible on a single picture. Only subsequent photos can focus on individual stains. It is a mistake to take pictures of the singular stain and ignore the documentation of how they are arranged in relation to each other. The keyword to help understand this algorithm is "from general to specific." This principle applies not only to the bloodstain pattern analysis but also to the documentation of the evidence.

Sometimes pattern that we, as investigators, are trying to document may be located far apart. The distance may be so great that it is impossible to fit them in one photo. In this case, it is a good practice to photograph the areas according to how we move. In such a case, at least one detail or an object linking the previous photo with the next should be shared in the background. Thanks to this, the person reading the material will not feel lost or teleported. Instead, it should feel some kind of continuity.

2.8 Science and Practice

Science and practice in bloodstain pattern analysis have almost fundamental importance (Beresford et al. 2020). Practitioners pose problems and questions. Scientists try to solve them and answer the unknowns. Science also indicates the limits and limits of the use of specific methods and technologies, so that practitioners know when to refrain from further work or inference. It is a cycle that drives and brakes itself at the same time. This allows for a relative balance.

In this book, I have tried to keep that spirit alive. At first glance, it may seem that relatively little space is devoted to the presentation of the basics of bloodstain pattern analysis. That information can be found in the extensive bibliography that I have attached to each chapter. It seems to me that scientific discourse lacks a discussion about mistakes. Errors that can be made while preparing reports in the field of bloodstain pattern analysis. That is why I place so much emphasis on discussing this area in the following chapters: homicide and suicide. Investigators and practitioners should know what can be determined using bloodstain pattern analysis and what is beyond the reach of this field of forensics. Without this information, the reconstruction of the event based on the bloodstains seems to be omnipotent and omniscient, which is not valid. This situation can be used by uneducated experts who will prey on the investigators' ignorance. With a tragic effect for suspects, accused and convicted base on opinions/reports in which there are errors.

If during the reading, the reader wants to learn more complex issues or wants to find out on what basis I present certain theses, it will not be a problem. As previously mentioned, each chapter contains a broad, comprehensive and most up-to-date bibliography. For those who are insightful and curious, it will be an excellent read and a source of additional knowledge, but not everyone can afford such detailed work. Sometimes it is enough for investigators to read the necessary literature only briefly. In such a case, they should have the opportunity to learn not only the basics but the most common mistakes that the practitioner sees in their work. Thanks to this, in the future it will be easier for them to verify whether a given report is written within the field or is an over-interpretation. This is another example of the coexistence of science and practice.

References

Al-Eid RA, Ramalingam S, Sundar C, Aldawsari M, Nooh N (2018) Detection of visually impercep-
  tible blood contamination in the oral surgical clinic using forensic luminol blood detection agent. J
  Int Soc Prevent Commun Dentistry 8(4):327–332. https://doi.org/10.4103/jispcd.JISPCD_10_18
Beresford DV, Stotesbury T, Langer SV, Illes M, Kyle CJ, Yamashita B (2020) Bridging the gap
  between academia and practice: perspectives from two large-scale and niche research projects in
  Canada. Sci Justice 60(1):95–98
Bevel T, Gardner RM (2008) Bloodstain Pattern Analysis with an Introduction to Crime Scene
  Reconstruction, CRC Press. FL, USA, Boca Raton
Camana F (2013) Determining the area of convergence in Bloodstain Pattern Analysis: a prob-
  abilistic approach. Forensic Sci Int 231(1–3):131–136. https://doi.org/10.1016/j.forsciint.2013.
  04.019
Chingthongkham P, Chomean S, Suppajariyawat P, Kaset C (2020) Enhancement of bloody finger-
  prints on non-porous surfaces using Lac dye (Laccifer lacca). Forensic Sci Int, 307. https://doi.
  org/10.1016/j.forsciint.2019.110119
Choromański K (2013) Bloodstain Pattern Analysis—Basic terminology. *Problemy współczesnej
  kryminalistyki nr 17, s.7–13;* ISSN: 1643-2207
Choromański K (2017) Using scientific method in bloodstain pattern analysis—goal of creating
  new and proper terminology *Przegląd badań z zakresu kryminalistyki i medycyny sądowej 31–38
  ((Wydawnictwo Naukowe TYGIEL sp. z o.o.: Lublin, 2017),* ISBN 9788365598479
Choromanski K (2020). Performing bloodstain pattern analysis and other forensic activities on cases
  related to coronavirus diseases (COVID-19). Int J Legal Stud 1(7):13–24
Conlan XA, Durdle A, Pearson C, Hayes R, Woolley Z, Stevenson PG (2019) Application of a
  digital stringing protocol on buried fabrics. Austr J Forensic Sci 51:S145
Ferrucci M, Doiron TD, Thompson RM, Jones JP, Freeman AJ, Neiman JA (2016) Dimensional
  review of scales for forensic photography. J Forensic Sci 61(2):509–519. https://doi.org/10.1111/
  1556-4029.12976
Gardner RM (2006) Defining a methodology for bloodstain pattern analysis. J Forensic Identification
  56(4):1
Gouse S, Karnam S, Girish HC, Murgod S (2018) Forensic photography: prospect through the lens.
  J Forensic Dental Sci 10(1):2–4. https://doi.org/10.4103/jfo.jfds_2_16
Hong S, Seo JY (2015) Chemical enhancement of fingermark in blood on thermal paper. Forensic
  Sci Int 257:379–384. https://doi.org/10.1016/j.forsciint.2015.10.011
James SH, Kish PE, Sutton TP (2005) Principles of bloodstain pattern analysis: theory and practice.
  CRC, Boca Raton, Fla
Matisoff M, Barksdale L (2012) Mathematical & statistical analysis of bloodstain pattern. Forensic
  Examiner 21(1):21–33
Larkin BJ, Banks C (2013) Bloodstain pattern analysis: looking at impacting blood from a different
  angle. Australian J Forensic Sci 45(1):85–102. https://doi.org/10.1080/00450618.2012.721134
Latham HM (2011) Using and articulating the scientific method in bloodstain pattern analysis. J
  Forensic Identification 61(5):487
Praska N, Langenburg G (2013) Reactions of latent prints exposed to blood. Forensic Sci Int
  224(1–3):51–58. https://doi.org/10.1016/j.forsciint.2012.10.027

Sterzik V, Panzer S, Apfelbacher M, Bohnert M (2016) Searching for biological traces on different materials using a forensic light source and infrared photography. Int J Legal Med 130(3):599. https://doi.org/10.1007/s00414-015-1283-2

SWGSTAIN, Scientific working group on bloodstain pattern analysis: recommended terminology, Forensic Sci. Commun. 11 (2009)

Chapter 3
Homicide Scenario

Abstract Homicide is an extremely difficult kind of procedure. It's usually very complicated, multithreaded and obscure. In this chapter, the author will try to break down the problems of applying bloodstain pattern analysis in such cases, based on a real-life example of a homicide case.

Keywords Spatter pattern · Victim · Reconstruction of the event · Examination of clothes · Crime scene investigation

Homicide is a particularly serious incident. It bears huge consequences for both society and the machinery of government. Police officers, forensic investigators and prosecutors are required to put in long hours, with impeccable work ethics and professional standards. The key to solving homicide cases is determining the detailed course of events and identifying its participants and their motives. Proving that the suspect committed the crime is of paramount importance. Such argumentation should be based on strong, scientific evidence and properly carried out interrogations. It is essential that the conclusions reached through the investigation of evidence point towards the perpetrator beyond the reasonable doubt.

An expert performing an analysis, including bloodstain pattern analysis, should realize the idea his function serves. His specialization should not lead to solving the case solely by proving someone's guilt. Ascertaining whether the suspect was involved in the crime or not is equally important. In a way, an expert's opinion may be compared to a double-edged sword. Science answers questions both through affirmation and refutation, and both are equally relevant and important.

During the homicide cases, law enforcement authorities bear a huge responsibility. In reality, they deal with at least two human lives, and each of them should be treated with an equal amount of respect—the life that was taken, and the life of a person that took it. On first glance, such a perspective doesn't seem all that intuitive. It is understandable that we, the society, want to make amends for the victim's death. To find and to punish his killer. However, as a society, we have matured. Nowadays, it is not just about finding a scapegoat, the spectacle of his punishment, fuelled by the blind thirst for vengeance. It is about proving his guilt, based on the analysis of

K. Choromanski, *Bloodstain Pattern Analysis in Crime Scenarios*,
SpringerBriefs in Forensic and Medical Bioinformatics,
https://doi.org/10.1007/978-981-33-4428-0_3

facts and scientific evidence, and then punishing him accordingly. Thanks to such an approach, chances of wrongly convicting an innocent person are being reduced to zero. These are the foundations of a just trial, that will benefit the entire society.

How are those values supported by bloodstain pattern analysis, and how can it be utilized during a criminal trial? How its results help investigators in their search for truth? In the following chapter, I will be discussing an actual homicide case, in an attempt to break down the problems one might face. I will discuss the course of the investigation, the traces secured by the investigators, and how they impacted the conclusions drawn. Furthermore, I will highlight the parts and nuances of the cases, which could have been overinterpreted or just misinterpreted by an expert, which would lead to errors in the following proceedings. Thanks to such an approach, I will be able to present scenarios, which are not often touched upon in specialized literature but are still, unfortunately, the bread and butter for the investigators. Showing the potential errors, mistakes and misconclusions may raise awareness and set clear boundaries for drawing conclusions in the field of bloodstain pattern analysis.

3.1   Introduction to the Case

In the late autumn morning, the police and an ambulance were summoned to a single-family house. Initial information received concerned the discovery of man's corpse. The person who called it in turned out to be the wife of the deceased. There were also two adult children of the woman and the deceased present in the house.

First statements from the interviewed family members indicated that it was a break-in and a homicide. Children of the deceased stated that they hadn't heard anything, as they were sleeping all night.

The initial version of events presented by the wife of the deceased was much richer in details. She stated that the whole family went to sleep in the evening. The children were to be sleeping in their rooms, and the husband in their bedroom. Children's rooms were on the same floor as the bedroom in which the husband was sleeping but on the opposite side of the hallway. One room was on the north-eastern side of the building, the other on the south-eastern side. The wife claimed that she went to sleep in a different room than her husband that night, the room located opposite of their bedroom. Which means that all family members were on the same storey of the building on the night of the husband's murder.

Further statements presented by the woman included information that might have had something to do with the event itself. She woke up in the middle of the night to a sound of glass breaking. However, she didn't get up from her bed. She didn't hear any further strange noises and she went back to sleep. Only in the morning did she realize that her husband wasn't responding to her calls. Disturbed by the silence she was to go into their bedroom and find her husband's corpse. While standing in the doorway, she turned around and went to the other room, where she made a call to notify the law enforcement authorities about the event. Then, she woke up her children and all three waited for the police and the ambulance to arrive. When the authorities arrived on site, the family was taken in for questioning. In the meantime, the crime scene investigators started inspecting the scene, both the house and its surroundings. The traces and objects secured by them started drawing a different scenario of the event.

3.2 Description of the Scene

The property in which the man's corpse was found was in between two smaller towns. The area around the house was surrounded by unpopulated farmlands and meadows. Several dozen metres to the north there was a narrow stream. The building nearest to the property was a landfill, half a kilometre to the north. It is safe to assume that the location was desolate. The only way to get to this area was by a scarcely used road, leading from north to south.

The building in which the event took place consisted of a warehouse and a residential part. The former was in the eastern part of the building, and the latter in the western. The residential part consisted of two storeys, with bedrooms on the first floor. The warehouse and a leisure area were on the ground floor. When the police arrived it turned out that the house alarm was disabled, and the local security force not summoned. There was no visible damage to the locks, gates, main or side doors. The only damage to the property that was discovered was a broken window in the utility room on the southern side of the ground floor, just below the room in which the wife was to be sleeping on the night of the event. Besides the broken glass, there was also a brick found in the utility room. Subsequent forensic research didn't reveal any fingerprints nor biological traces on the brick. The investigators looked for a place from which it might have been taken. For that, the entire property and its surrounding were inspected. Nearby the broken window, within a few metres from the southern wall of the building, investigators found some broken glass and ceramics, alongside discarded placemats. However, nothing that could be connected to the brick was found there. As it turned out, the solution to this puzzle was closer than the investigators have thought. On the northern side of the house, by the garage door, there was a carpet lying on the ground. Most likely it was supposed to work as a sort of a doormat. It was being held down by three identical bricks—one on each corner of the rug, and each one was exactly the same as the one found inside. One of the rug's corners was not being held down by a brick. There was, however, an indentation the size of a brick on it. It leads the investigators to believe that this is where the brick, which was most likely used to break the window, was found.

Further investigation of the broken window some more information. There were pieces of broken glass on the window sill, which was no surprise. What was interesting, however, was that there were no boot prints on the sill, even though the area around the building was wet and marshy.

There were also several outbuildings on the property, besides the main residential building. They were also investigated by the police, however, none of them showed any signs of entry. It's worth noting that in one of those buildings they found a microwave oven that was smashed against the bricks.

There were some traseological traces found on the property, in a form of indentations in the ground. Unfortunately, the poor quality of these traces and their lack of distinctive properties didn't allow the subsequent forensic traseology report to identify neither the footwear nor the person.

Outside of the property itself, the police was able to secure a very interesting object. A black wrapped-up plastic bag was found in the shallow water. It contained an axe and a towel with dark-red traces on it. These objects were secured for fingerprint identification. Later on, they were to be tested for DNA profiling, as a part of a forensic DNA identification. Detailed conclusions that resulted from these tests will be presented in the following subchapters.

As it was previously mentioned, the only traces that could've been connected to the break-in were found in one of the utility rooms, on the ground floor of the residential building. The corpse of the deceased was found on the first and highest floor, of the northern side of that building, just by the stairs. The victim was discovered in his bed, lying down. There were multiple bloodstains in this room, which will be described in detail in the following subchapter.

There is one thing that should be noted. The victim had a firearms licence, in this case, it was a hunting rifle. In Poland, it is a somewhat rare occurrence, since firearms accessibility and its possession is not as common as in, for example, some states in the USA. Said rifle was found under the victim's bed. It was loaded, but the safety was on.

3.3 Bloodstain Pattern Analysis at Crime Scene

As it was previously stated, the crime scene investigation covered both the residential building and the nearby buildings. The only red and russet stains that were suspected to possibly be blood were found in the room in with the body. It was a bedroom with a single large bed, two bedside tables, and windows facing north. There were white laced curtains by the windows and individual paintings on the walls. The floor under the bed was covered by a carpet. The general layout of the room is best shown by Fig. 3.1 photo.

On the wall to the right from the entrance, beneath a souvenir photo, eleven individual red and russet stains were found, ranging in size from 0.8 mm to 1.5 mm. Their location is presented in Fig. 3.2 photo. The close-up of one of the stains is presented in Fig. 3.3 photo. All of these stains had clearly distinct edges. The tails of stains were facing down, indicating that the drop that had made them was moving in a downward direction. Unfortunately, it was impossible to determine whether that was due to the source of the spatter being above them, or it was below and the drop had made a parabolic trajectory before making contact with the wall. If the tails

Fig. 3.1 The general layout of the room in which the man's corpse was found

Fig. 3.2 The right wall of the room in which the man's body was found, as seen from the entrance. Beneath the picture four red and russet spatters are located

Fig. 3.3 A russet and red spatter, with its tail facing down, located on the right wall of the room in which the man's body was found

of stains were facing up you could categorically state that the source of the spatter was located below the stain, and the drop was moving in an upwards direction, thus resulting in such a positioning of the tail (Kunz et al. 2014). In Fig. 3.3 photo there is no established base line, a line perpendicular to the direction of gravitational force. Without such an indication it is impossible to determine the stain's angle from the photo, rendering successful concluding impossible.

The man's corpse was discovered in bed. He was found in a position presented in Fig. 3.1 photo. Figure 3.4 photo presents a close-up of his head and back. There are numerous russet and red spatters on the pillows, headboard, and sheet. There are nearly no stains on the victim's back. There are individual contact stains on the victim's shoulders. Spatter patterns on the pillows and the victim's hands do not overlap. There are no visible contact stains on the victim's hands nor face. The flow pattern, presented in Fig. 3.5 photo, found on the left side of the victim's face is facing downwards.

On the left wall, as seen from the entrance, there were numerous russet and red spatter patterns. The spatters were oval-shaped, ranging in size from 0.5 to 2 cm. There was also a stain from which an intense, dozen centimetre long low pattern derived in a downward direction. Unfortunately, there are no detailed pictures of this stain. The russet and red spatters on the left wall by the window were located on its all height, as shown in Fig. 3.6 photo.

Fig. 3.4 Photo presents the size, shape, location, and mutual placement of spatter patterns found on the victim's body, pillows, sheet, and headboard. The photo was taken from the right side of the bed, as seen from the entrance

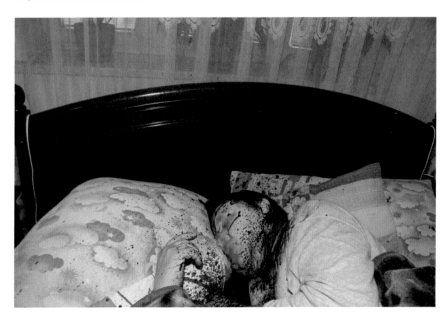

Fig. 3.5 Photo presents the size, shape, location, and mutual placement of spatter patterns found on the victim's body, pillows, sheet, and headboard. The photo was taken from the left side of the bed, as seen from the entrance. Face of the victim is blurred for anonymity

Fig. 3.6 Photo showing the placement and location of numerous russet and red spatter patterns found on the left wall and the curtain on the front wall

Besides the spatter patterns on the left wall, there were also saturation patterns found on the curtain. Considering their location, it is possible that the original trace was a spatter that seeped into the surface, spread across the fabric, and turned into a saturation pattern (Fig. 3.7).

Spatter patterns' edges were clearly distinct and their tails were facing downwards. Unfortunately, there were no detailed photos taken during the investigation, rendering precise concluding impossible. In this situation using 3D scanner for determination possible cast off would be useful (Liscio et al. 2020), but during this time this software wasn't available yet.

A more detailed investigation of the bed allowed for a better understanding of how certain stains had been created. After the body and the quilt were removed from the bed, the investigators found the stains visible in Fig. 3.8 photo. They were saturation patterns, that seeped into the sheet, mattress, soaked straight through them, and kept flowing downward, resulting in a pool pattern with a visible bloody clot, as shown in Fig. 3.9 photo. There were also other stains on the bed. Russet and red spatter patterns on the headboard were extremely interesting. The spatter patterns were so numerous that they merged, forming a downward moving flow pattern. There were also contact stains on the sheet, as well as individual drips. It should be noted that these stains were not found or documented before the body was moved, thus it can be assumed that they were a so-called "artifact", a stain created after the criminal incident. In this case, they were created as a result of the investigators' work.

Fig. 3.7 Close-up photo showing placement and location of numerous russet and red spatter patterns found on the left wall and the curtain on the front wall, possible cessation cast of

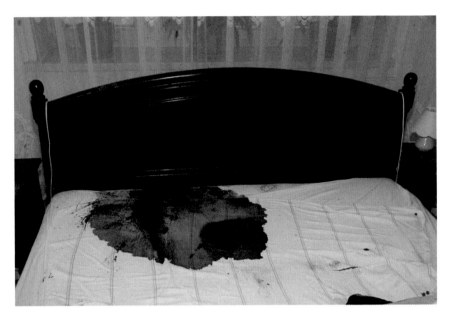

Fig. 3.8 Numerous russet and red spatters on the headboard with noticeable downwards flow patterns, as well as a visible saturation pattern on the mattress, creating a pool of blood

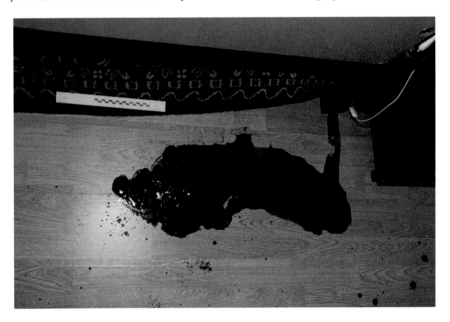

Fig. 3.9 Russet and red pool with a noticeable clot and a visible serum separation. There is a partial traseological print visible by the pool, as well as russet and red spatters and drips

Another observation worth mentioning is the presence of what seemingly appears to be a dilution by the pool on the trace presented in Fig. 4.9 photo. It is a so-called "serum stain," created as a result of the separation of blood's cellular components (the red part) from the serum (the bright fluid.) The biochemical and physicochemical reasons for this phenomenon are not fundamental from the analyst's point of view, however, they are an object of interest for some scientists.

You may have noticed that the pool of blood's upper part is exceptionally regular. The reason for it was a spatial obstruction, specifically the red carpet, that stopped the fluid from freely spreading. The edge of the carpet had stopped the stain's spread, resulting in its a regular shape. Once the bed and the carpet were moved, the Fig. 3.9 photo documented the end result.

It's worth noting that the pool of blood was not in its dried up form during the crime scene investigation, it had a form of a semi-liquid gel, which made the proceedings significantly more difficult. Such traces can easily be transferred, moved, or altered as a result of an inattentive work on the crime scene.

3.4 Bloodstain Pattern Analysis at Evidence

In this case, the police officers secured cloths that people residing within the house had worn, when the incident occurred, i.e. clothes of the wife and the children of the deceased. The police also secured the rug and the axe that they had found wrapped-up in the plastic bag, outside of the residential house, but within the boundaries of the property. For this case, the inspection of clothes was crucial for its solving. In this subchapter, the detailed information gathered from secured evidence will be described.

The evidence from the crime scene and the residents of the house were properly secured and sealed in sterile packaging. They were transported intactly to a genetic laboratory, for further, more detailed, analysis.

The Fig. 3.10 photo presents top of the pyjamas the wife of the victim wore at the time of the police's arrival. It's a button-front shirt made of cotton. The buttons are made of white plastic, and the fabric is uniform, but covered in a dappled pattern, consisting of small red and pink flowers on russet branches, with green leaves. The pattern is complex, fine and colourful, thus making its inspection a difficult task. To accomplish it, the analyst, alongside a group of lab workers, spent many hours meticulously inspecting the shirt. During those tests, they were able to reveal numerous russet and red stains, most of which were found at chest height and on the left sleeve. The stains found on the chest part of the shirt were located on a surface of 30 cm by 30 cm. The stains revealed on the left sleeve were located on nearly its entire length.

The Fig. 3.11. photo presents a close-up of a front of the shirt, at the chest height, by the sleeve, to help with the visualization of the stains. To facilitate finding them even further, the arrows were placed, pointing towards the individual stains. The spots vary in size, from 2 mm to less than 1 mm in diameter, and they're round or oval-shaped. Their edges are clearly distinct, and they are strongly saturated. The

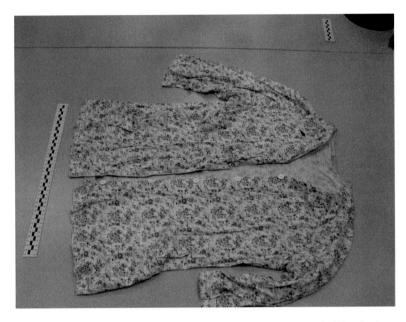

Fig. 3.10 The pink pyjama shirt with a dappled pattern, hindering the reveal of bloodstain patterns

Fig. 3.11 Part of the pink pyjama shirt with a dappled pattern, hindering the reveal of bloodstains

larger stains, visible to the left of the white arrow on a black background, vary in
diameter from 1.5 mm to 3 mm, and they're also saturated with distinct edges. There
are no signs of dilution or abrasion on the aforementioned stains.

The Fig. 3.11 photo demonstrates in close-up the problematic fabric, on which
the bloodstains had to be revealed and secured. The pink and red, relatively irregular,
dappled pattern made the work much more difficult, especially when you consider
the fact that some stains are less than 1 mm in diameter wide. Nevertheless, the
analysts managed to locate, document and take samples of the bloodstains, which
underwent genetic testing.

The Fig. 3.12 photo presents the inner side of the shirt, which is more uniform
because the pattern is less noticeable. Flipping the shirt inside-out made the location
of stains, that soaked through the material, easier and faster to trace. Nonetheless,
every stain had to be verified, whether it was created by transmission or as the result
of saturation. Such an analysis requires the use of a microscope, and is a part of a
more advanced battery of tests, from the field of bloodstain pattern analysis.

The following pictures (Figs. 3.13 and 3.14) present show pyjama trousers, worn
by the victim's wife. The trousers are bright coloured, uniform and not patterned.
They're made of cotton, with visible signs of usage, however, they're not worn.
Hence, the fabric is uniform, not just colouristically, but also structurally. There
are no visible holes or damages. The Fig. 3.13 photo shows the front side of the
trousers. There are noticeable russet and red stains on the left leg that were tested for
human blood. The test was positive. The photo was taken in a wide angle, in order

Fig. 3.12 Saturation stains as seen from the inside of the shirt, allowing for easier and faster tracing
of bloodstains

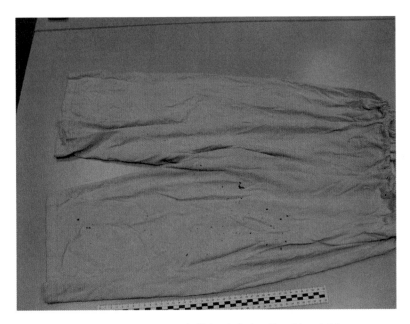

Fig. 3.13 The pink pyjama trousers, secured off the victim's wife, with a visible spatter pattern on the front side of the left leg, and individual spatters on the front side of the right leg

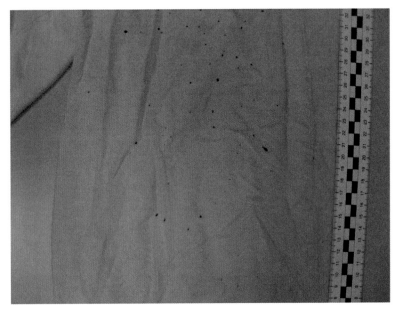

Fig. 3.14 Pink pyjama trousers, secured from the victim's wife, with a visible spatter pattern on the front side of its left leg

to thoroughly document the location and mutual placement of stains revealed during the inspection.

The Fig. 3.14 photo show the close-up of the trousers' left leg front side. The surface is covered with visible russet and red, round and oval-shaped stains. They are strongly saturated, and their edges are clearly distinct. They vary in size from 3 mm to less than 1 mm in diameter. They are located on a stretch of material that is 25 cm wide and 40 cm long. Similarly to the shirt, the fabric on which the stains were revealed is absorbent. Because of that fact, the analyst couldn't estimate the angle of impact or determine the area of convergence, or the area of origin. Performing detailed analysis on this kind of surface can be unreliable due to nature of surface (Taylor et al. 2016), so analysis wasn't so detailed. Unlike the previously discussed garment, the trousers are uniform and bright-coloured. The stains are easily noticeable on such surface, therefore, there is no need for arrows or pointers.

The Fig. 3.15 photo shows the axe with a wooden handle that was wrapped up in the rag, found within the plastic bag outside of the property. There are no visible russet and red stains on the tool. The axe head is made out of metal, with noticeable traces of rust and dirt. At first glance, the object appeared to have nothing to do with the event, as it appeared to be clean. That assumption is wrong. The true nature of the axe is visible at the Fig. 3.16. The blue glow is caused by the reaction between luminol and the substance present on axe's surface. One of the substances that cause such a reaction is blood, which contains haemoglobin, which contains iron ion. It has to be noted that rust causes a similar reaction. Which is why it is necessary to conduct further, more precise tests, after obtaining a positive result with forensic

Fig. 3.15 The secured axe in a natural light, with no visible bloodstains

Fig. 3.16 The secured axe in a natural light, covered in blue glow indicating a positive result of a luminol test

chemiluminescence, in order to confirm, whether the substance is human blood, and if so, to whom does it belong. It's worth notice that luminol have vast range of false positive effects (Quickenden and Creamer 2001; Ussher et al. 2009; Sorum 2013; Stoica et al. 2016).

3.5 Reconstruction of Event

While performing this case's reconstruction of the event, you have to highlight a single, extremely important matter. No documentation is perfect. There will always be some mistakes, errors, inaccuracies both in the techniques used to reveal the stains and in their further documentation. It sometimes happens that the investigators miss details crucial to the case's solving, and only realize that few years later. Almost always bloodstain pattern analyst operates with a limited material. You have to realize that fact and adjust your expectations regarding the expert. In this case, only the representative of the prosecutor's office and forensic investigators were present on the crime scene. There was no bloodstain pattern analyst, which is why some of the aforementioned photos contain slight technical inaccuracies. They are by no means catastrophic enough to render the analysis impossible, nevertheless, they have to be acknowledged.

The bloodstain pattern analyst working this case operated based on procedural documents, i.e. the crime scene investigation report, and photos taken during the inspection. Furthermore, the analyst took part in the inspection of objects in the genetic laboratory. During the tests, bloodstain patterns were revealed and documented, through detailed photos and descriptions. The material prepared in such a manner makes for an excellent base when presenting opinions and conclusions.

While performing the reconstruction of the events, the analyst based his conclusions on documentation prepared by other people (Choromański 2015). This situation doesn't match what you might see on the television, however, this is how the situation presents itself in reality. In the case of countries in which bloodstain pattern analysis is underdeveloped, the forensic experts are scarce, with new cases constantly flowing in, which makes this a common occurrence. Which is precisely why sharing knowledge and basics of bloodstain pattern analysis is so important.

It also must be noted that the bloodstain pattern analyst, while recreating the event based on the bloodstain patterns and drafting his report, was relying solely on physical evidence. He hadn't acquainted himself with witness' reports, testimonies, or police notes. His opinion was based only on crime scene investigation report, an inspection of clothes and objects, genetic test report and photos taken. Thanks to that, the analyst was able to limit the bias and based his opinion solely on scientific evidence. The documentation he used, allowed him to draw some interesting conclusions, which proved to be very helpful for the prosecutor.

The forensic medic's opinion indicated that the only wounds of the victim were cut and crushed, located on his head and neck. The victim was lying still in bed, the moment the blows were dealt, which is indicated by spatter patterns on headboard and bedsheet. The spatters are focused around a single area. The victim received these blows and did not move from where he was, which is indicated by the lack of wipes of existing patterns around the victim. The location of scatter patterns on the victim's hands, in regards to the patterns on the bedsheet, indicates that they also did not move. The assailant or assailants swung the murder weapon while dealing blows, which created numerous spatter patterns on the walls to the left and right of the entrance. The spatter patterns are cast-off spatters, with possible impact spatter patterns. Cesesion cast-off spatters cannot be excluded. The objects lying on the floor, on the left side of the bed, i.e. glasses and remote controls, were already in these places when the russet and red spatters on the floor were created, as indicated by spatter patterns of similar shapes and sizes that were found on these objects. The pool of blood, located under the bed, on the floor, by the window wall, is most likely caused by blood soaking through the mattress, as indicated by saturation stain on the mattress. The russet and red traseological traces, found between the bed and the window wall, were most likely created by the technicians or a medical examiner, as indicated by the lack of these traces on the photos taken before the corpse and bed were moved. Further tests were conducted on the axe after the stains that could have been blood were revealed. The results have shown that the glow was indeed caused by the presence of human blood, which matched the victim's genetic profile. The axe made contact with the blood, as indicated by numerous bloodstains of the victim covering the weapon. Unfortunately, the details of how these stains were created is impossible to determine, as they are not distinct enough. The person wearing the pink pyjamas was in the vicinity of the area of origin, as indicated by spatter patterns on the left sleeve, chest and left leg. The location of these spatters indicates that it's probable that the person wearing the garment was near the area of the origin of spatter as a witness or assailant or cooporative.

3.6   Possible Areas for Mistakes

In this subchapter, I will try to touch upon some of the areas in which a bloodstain pattern analyst may make mistakes. The consequences of different stain localization and the possibility of its transfer." The problems described in this chapter will concern the matters of drawing conclusions based on bloodstain patterns, and their possible overinterpretation. Its goal is to show the broader picture of bloodstain pattern analysis.

The first very serious mistake, while drawing inferences in this case, might concern making a statement that the victim was hit while he was sleeping. Many clues are supporting that conclusion. The spatter found on the bed frame unequivocally implicates that the source of the spatter was on the mattress. Considering the placement of wounds on the victim (skull fracture), you can determine that the person lying on the bed received a blow to the head, while his head was placed on the pillows right by the headboard. This is, however, all that you can determine from that fact. There is no evidence pointing towards the conclusion that the victim was sleeping while the blow fell. Indeed, the source of the spatter was not moving, but that doesn't mean he was asleep.

On first glance, such a mistake doesn't seem that relevant. After all, what's the difference, whether the victim was sleeping or simply lying in bed? In fact, it might make a huge difference. First and foremost, drawing conclusions based on incomplete knowledge is unethical and unscientific. It is not the role of an expert, to present himself as the alpha and the omega. An expert does not have to know everything. This is not a TV show, where people in the lab know every answer to every single question. An expert must realize his or her limits and cognitive boundaries. As a scientist, you make assumptions based on facts, not presumptions. If you are unable to determine something, there is no shame in that; you just write that the given state, event or fact is impossible to verify. Second of all, the legal classification of an attack on a conscious person lying in bed may be different than one of an attack on a sleeping person, unaware of his assailant. Which means that an expert presenting a specific version of the events, may indirectly shape the future indictment, or even the sentence. Which is why it is absolutely unethical and reprehensible to overinterpret traces in that manner.

The second commonly occurring conclusion in such cases, which is unfortunately incorrect, is presenting a theory that the event was an act of aggression against the victim. It is a fact that in this case, the victim received multiple blows with an axe to the area around his head. Bloodstain pattern analysis, both on the scene and the clothes secured from the scene, does not allow for a statement regarding aggression or the assailant's emotional state. BPA may determine that source of the spatters was on the bed while the blows landed. In some cases, it is possible to determine the minimum the number of hits. However, these pieces of information are not directly connected to the emotional state of the assailant. The attacker may land few hits while being either calm or aggressive, and the wounds will look exactly the same, since it is related to the force of the blow, and where they landed, not with the feelings of the assailant. Therefore, if a bloodstain pattern analyst states in his opinion that the person involved in the event was aggressive, angry, furious, or calm, it is clear that it's a huge overinterpretation, not supported by any evidence. Such conclusions may be important for the prosecutor, for whom the information concerning aggression will prove handy since it will strengthen his indictment. However, you have to remember that this information is not supported by anything, and it can be shaken by the defence when the case goes to court.

Another conclusion that can be mistakenly extrapolated from gathered evidence concerns the towel and the axe, wrapped up in the plastic bag. A careless bloodstain pattern analyst may deduce that this object was hidden or purposefully thrown away. However, you have to ask yourself whether the evidence actually implicate that fact. Which stain exactly inform the analyst that the object was abandoned in order to be hidden? The answer may be surprising, but there are no traces that point towards that conclusion. The only traces that were revealed in this case, were diluted stains. The only thing you can tell from analyzing them is that the stain was mixed with another substance. They say nothing of the blame or the state of mind of a person that wrapped the object in a rag and left it in a plastic bag. In an older Polish literature, such traces would be classified as washed stains, which is, in my opinion, an error in categorization. Additional area for overinterpretation you might come across, concerns making far-reaching conclusions based on bloodstains revealed with chemiluminescence. The stain may be invisible due to its dilution, or a weak contrast between the trace and the surface on which it was found. The fact that the stain glows, when forensic chemiluminescence is applied, does not indicate that the object was cleaned or washed. Currently, this topic is being researched by scientists. But let's come back to the subject of analyzing and drawing conclusions based on stains revealed on the rag and the axe found within the bag. It's not for the bloodstain pattern analyst to indicate a guilty party, or to "enter the mind of the criminal." The analyst's only duty is to objectively recreate the facts, based on the evidence. It's the responsibility of the prosecutor and the judge to make something of these facts and possibly connect them to the crime. Unfortunately, it happens that in many cases the analysts forget about that fact, which often has detrimental consequences.

Another possible area for mistakes, which could be made in this case, is a wrong identification of one of the bloodstains. The stain in question was a clotted pool of blood revealed on the floor, under the bed, with a noticeable yellowish stain. For an untrained eye, this stain could be miscategorized as a diluted stain, which would explain the presence of this odd substance. That would be an incorrect assumption. The substance's source is the blood itself. It is serum, that was created when some of the blood separated from its cellular components, which are the red part of the stain. This process is called serum separation. As it was said before, it may be mistaken with stain dilution, which is a serious mistake. Currently, researches are being conducted, that would help the analyst determine the time of the bloodstain's creation, by correlating how the dry pool of blood looks with the timetable of its drying and coagulation.

In the room in which the corpse was found, it was possible to reveal the russet-red cast-off spatters on the walls. According to some authors, it is possible to determine based on these traces whether the person taking swings was left or right-handed. Other authors, however, state that this information is impossible to determine. Since in my professional work I try to categorically point out relevant facts, based on scientific data, I gravitate towards the latter opinion. That's why in this, and other cases, I don't determine based on cast-off spatters whether the person was right or left-handed, since currently the research results on that matter aren't unequivocal enough, and there are many unknown factors. I'll try to briefly explain what makes it so complex. It's one thing to determine on what plane, or planes, the murder weapon was moving while the hits were being dealt. It's a whole other thing to determine the handedness of the attacker. While using some tools as weapons, especially one-handed tools, a certain range of movements is much easier to execute by a right-handed person, and the same goes for a left-handed person. But this is not a rule. Hitting someone is an extremely dynamic occurrence, which is influenced not only by the assailant's handedness, but also by other factors, such as the position and location of the object hit, the position of the hitter, the weight of the tool, and whether the attacker uses one hand or both. Moreover, all sorts of spatial obstructions, such as walls, furniture and other objects that inhibit or facilitate the swings also play a part. Only such a detailed description of a situation allows for a fuller understanding of the nuances of drawing conclusions. In the discussed case you can determine the planes in which the swings were taken, however, further speculation may lead to overinterpretation.

The next mistake that could've been made in this case was an overinterpretation of spatter patterns found on the secured nightgown of the victim's wife. An inexperienced analyst could state that the spatter categorically and unequivocally indicates, that it was the wife who dealt the blows. But is it a correct assessment, or is it too categorical? Does the fact that the scatter patterns are found on someone immediately indicates that this person was the assailant? No. The presence of spatter patterns on any given surface only indicates that said surface was nearby the area of origin. In this case, the spatter localization indicates that the leg, left arm and chest of the gown was in the vicinity of the area of origin. Such spatter localization indicates a high degree of probability, that the person wearing the gown, when the spatter was created, was the one that dealt the blows. Still, however, it is only a high degree of probability, not a proven fact.

There are situations, in which traces found on a sleeve of clothing may indicate that this garment was in the vicinity of the impact spatter pattern, but it still doesn't directly indicate that the person wearing it at the time the spatter was made, was the one dealing the blow, and, therefore, guilty of the attack. These nuances at first may seem insignificant. Said information and the conclusions based on them may influence the legal classification of the act, resulting in a different sentence. Given example presents a situation, in which the analyst used a mental shortcut, because if there are spatter patterns, then you can categorically state that this person must've been the attacker, right? Not necessarily. That's why the analysts, the prosecutors and the judges alike should be sensitive to these kinds of statements, and within the context of this case, the question "whether the person wearing the gown was the attacker?" was answered only with the fact that it was "highly probable."

The next mistake that I would like to describe within the context of this case, is specifically pointing towards the murder weapon, based on the impact spatter pattern. Actually, there is no scientific basis that allows the analyst to determine what specific tool was used to create any given impact spatter pattern. It was showed in one of the popular TV series, but for now, there is no scientific data that would allow for such a leap. Due to numerous questions, I've received from fellow practitioners, and people commissioning BPA opinions, I have decided to include that piece of information in this book, to impart the most current knowledge in the area of bloodstain patterns based reconstruction of events.

Another area for mistakes, in this case, concerns estimating the number of blows dealt, solely based on scatter patterns found on the headboard. It is true that during the crime scene investigation, some impact spatter patterns with flows of blood were found. However, based on their appearance and mutual localization, they were created almost simultaneously. In certain places, the spatters overlap, resulting in larger spatters and creation of blood flows. Flows can also be created as a result of the application of a relatively large amount of blood on the surface. Furthermore, the area of origin's localization indicates that it resided on the mattress, by the headboard. Overlapping spatter patterns make the number of blows dealt impossible to determine. However, you can imagine a different scenario. If the area of origin moved in between the blows, the spatters would also appear in other locations. In an ideal scenario, they would be so far from one another, that you could easily pinpoint the exact locations. In such a hypothetical scenario, it would be possible to estimate the minimal amount of blows, based on clearly distinguished and properly analyzed impact spatter patterns. Unfortunately, in the given scenario the spatters overlap, which means that the only thing the analyst may definitively deduce, is that at least two separate hits were the source of the spatter patterns. Such information, however, is also beneficial to the case. To categorically determine the minimal amount of blows dealt, the bloodstain pattern analyst should combine the analysis of impact spatter patterns with the medical examiner's report, or the opinion of a mechanoscopy expert.

3.7 Conclusion

Homicide has been committed in a typical family house. We could say that this is a classic situation for this type of cases. Additionally, the participants of the event are in the so-called family circle. In this case, the victim's wife was indicted and subsequently convicted of her husband's murder. How did the bloodstain pattern analysis help in this case? It was possible to reconstruct the event and determine its course without BPA. It was possible to indicate which surfaces were near the impact spatter when hitting the blood occurred. In this case, the source of the blood was the head of the victim. Could this information be obtained without using bloodstain analysis? Perhaps it would have been possible to acquire detailed information on the course if the wife had told precisely what she did and what she had invented for the purposes of the proceedings. However, it seems more logical to use science to verify forensics than to take anyone's word for it.

In fact, there were minor mistakes in this case, but nevertheless, those errors did not prevent reconstruction of the event. Reconstruction was not categorical and extremely detailed. But it made it possible to come to the truth. Besides, it made it possible to exclude other versions that were presented during the proceedings. In my opinion, this is an excellent example of a case where properly applied blood mark analysis facilitates the conduct of proceedings. Thanks to it, the matter becomes even more transparent and easier to understand. This isn't first time when bloodstain pattern analysis is helping investigator to determine is the event suicide or homicide (Pelletti et al. 2017). But it's worth putting next example into the field.

References

Choromański K (2015) Bloodstain pattern analysis—practical aspects of the forensic discipline. Edukacja Prawnicza Issue 161(5):23–26; ISSN: 1231-0336

Kunz SN, Klawonn T, Grove C. (n.d.) Possibilities and limitations of forensic bloodstain pattern analysis. Wiener Medizinische Wochenschrift 164(17–18):358–362. https://doi-1org-10000b5r2 07ed.han.buw.uw.edu.pl/10.1007/s10354-014-0297-6

Liscio E, Bozek P, Guryn H, Le Q (2020) Observations and 3D analysis of controlled cast-off stains. J Forensic Sci 65(4):1128–1140. https://doi.org/10.1111/1556-4029.14301

Quickenden TI, Creamer JI (2001) A study of common interferences with the forensic luminol test for blood. Luminescence: J Biolog Chem Luminescence 16(4):295–298

Pelletti G, Visentin S, Rago C, Cecchetto G, Montisci M (2017) Alteration of the Death Scene After Self-stabbing: a case of sharp force suicide disguised by the victim as a homicide? J Forensic Sci 62(5):1395

Sorum ED (2013) Identifying a false positive reaction from Bluestar on nonporous surfaces. J Forensic Identification 63(6):660

Stoica BA, Bunescu S, Neamtu A, Bulgaru ID, Foia L, Botnariu EG (2016) Improving luminol blood detection in forensics. J Forensic Sci 61(5):1331

Taylor MC, Laber TL, Kish PE, Owens G, Osborne NKP (2016) The reliability of pattern classification in bloodstain pattern analysis—PART 2: bloodstain patterns on fabric surfaces. J Forensic
Sci 61(6):1461

Ussher SJ, Milne A, Landing WM, Attiq-ur-Rehman K, Séguret MJM, Holland T, Achterberg EP,
Nabi A, Worsfold PJ (2009) Investigation of iron(III) reduction and trace metal interferences in the
determination of dissolved iron in seawater using flow injection with luminol chemiluminescence
detection. Anal Chim Acta 652(1):259–265. https://doi.org/10.1016/j.aca.2009.06.01

Chapter 4
Suicide Scenario

Abstract The true understanding of what actually happened in suicide cases is crucial for at least two reasons. First of all, the analysis and determination of course of an event allows for the exclusion of a third party. Second of all, and equally important, is the recreation of the scenario that resulted in a person's death is a moral duty to the family of the deceased and the community. It forms a sort of a frame. A conclusion, allowing for a full understanding of what happened and possibly allowing to prevent such future events. In this chapter, a case that had happened in a very peculiar place and setting will be presented. Despite these differences, certain rules and principles apply in the same way as they do in homicide cases. The author hopes that the presentation of the analysis, conducted during this case, will convince the reader that the application of bloodstain pattern analysis should be a standard procedure.

Keywords Suicide · Razor blade · Spatter pattern · Flow pattern · Contact pattern

Suicide is a very difficult situation. It burdens not only the family of the deceased but also it carries over to the society and the machinery of government, represented by the prosecutor's office and police. Discovery of a corpse carries serious consequences and forces law enforcement to take certain actions. The priority is establishing which of the four possible scenarios happened. Was the death caused by a third party, was it an accident, was it of natural causes? The fourth possibility is suicide. The last two options carry serious legal ramifications in the light of the criminal law. In certain regions of the world, suicide is penalized, however mostly on religious grounds, or in civil cases, concerning inheritance issues. Instigating to commit suicide can be punished, however, it is not a concern for bloodstain pattern analysts. It does not change the fact that distinguishing suicide form homicide or staged suicide is crucial for criminal proceedings. If one of the scenarios can be ruled out from the start of the investigation, the investigators will save time and money. An investigation leading to a dead-end is pointless. Application of bloodstain pattern analysis can save the investigators from taking inefficient actions.

Not all suicide cases present with intense exsanguination. However, when they do, a bloodstain pattern analyst should take part in it.

© The Author(s), under exclusive licence to Springer Nature Singapore Pte Ltd. 2020   41
K. Choromanski, *Bloodstain Pattern Analysis in Crime Scenarios*,
SpringerBriefs in Forensic and Medical Bioinformatics,
https://doi.org/10.1007/978-981-33-4428-0_4

4.1   Introduction to the Case

The corpse of a 50-year-old woman was found on a summer morning in a very peculiar place. The victim was found in a restroom. There were numerous russet and red stains all over the room. People found in a neighbouring common room were detained and isolated from each other. Almost immediately after she was found the police and the prosecutor's office were notified, and law enforcement authorities soon arrived on the scene, as well as an expert medical examiner. After the victim was officially pronounced dead, the scene was secured, so no one could interfere with the way it looked. When the prosecutor in charge of the case received the initial report, he deputed a bloodstain pattern analyst to take part in the investigation and secure evidence.

4.2   Description of the Scene

There would be nothing extraordinary in this event if it weren't for a fact that it took place in prison. This particular setting forces very specific working conditions upon the investigators. In this case, they were extremely favourable. The investigators were dealing with a locked-down, and somewhat sterile crime scene, where the number of people capable of committing the crime was limited. In this case, it was limited to the adjoining cell, in which three other women resided, and the restroom in which the body was found. The cell was leading out to a hall, which was constantly monitored. These limitations made the proceedings much easier. However, it doesn't change the fact that meticulous and complex investigation had to be conducted.

The cell was built to contain five people, however, during the events, it was occupied by four people, including the victim. It's worth noting that the cell was destined for people in provisional detention, which is why the presence of numerous personal items comes as no surprise. The security regime in such cells is different from regular cells for convicted prisoners. The room contained two bunk beds and one regular bed. There were personal cupboards under the beds and on the walls, containing women's personal belongings. The cell also contained several stools and a table, on which food and personal belongings, such as glasses or handkerchiefs were found. The floor was tiled to make the surface easier to maintain. In the wall opposite the entrance was a barred window, which was open when the examination commenced, most likely due to the high temperature in the room, even at this early hour. As was mentioned, the room contained numerous personal items of the detainees, such as clothing, shoes, medicine, dietary supplements, dry food, simple jewelry, and toiletries. The investigators found the room in a reasonably well-maintained state. There was no evidence of attempts to scour through the detainees' belongings. The fact that the victim's body was found in the morning supports the hypothesis that the detainees had no time to hide anything in their cupboards.

During the examination of the cell, the investigators managed to find the cupboard with the victim's belongings. Except for the aforementioned personal items, the correspondence between the victim and her family was found. Due to the specific details contained within those letters, their contents will not be divulged. Furthermore, a single, used-up disposable razor was found; only its handle and a partially broken top. The cutting part was missing.

4.3 Bloodstain Pattern Analysis at Crime Scene

As it was previously stated, the russet and red stains were found in two rooms; the toilet and the common room. The toilet door was open wide. As the investigation commenced, the woman's body was in the common room, by the cell door. There were numerous russet and red stains found on the walls, floor, and objects of the restroom. The layout of the room is presented in Fig. 4.1 photo.

On the left wall of the restroom, as seen from the entrance, 80 cm above the floor, numerous russet and red oval-shaped stains were found. Most of them had their tails pointing downwards, however, some of them were pointing to the side. Their edges were clearly distinct, and their shape made them easy to identify (Gardner et al. 2012). Their sizes ranged from 1 mm to 2 cm. The amount of substance was sufficient enough to make it streak down the wall, creating flow patterns that make for the majority of the stain size-wise. These traces are documented in Fig. 4.2 photo. Furthermore, the traces had been wiped; a contact stain was found under the roll of toilet paper. It was created as a result of contact of the roll with the wall's surface, as the stains were still not dried up. The movement of the another surface (could be hand) caused the change of preexisting spatter and low patterns.

Spatter patterns were also found on the wall, right by the floor. However, their shape and size differ from those that were found above them. The analysts were able to determine that these stains were created through two overlapping mechanisms. The first is the same one that created the stains above, while the second was related to the substance dripping onto the floor, causing spatter patterns on the wall.

The russet and red, oval, and round stains were found on the floor of the restroom. Their size ranged from 1 cm to less than 1 mm. The stains cover the majority of the room's floor and were categorized as spatter patterns. Considering their location, shape, and the fact that a pool, that was possibly their source, was found in their vicinity, the investigators were able to deduce that the spatter patterns were created as a result of substance dripping and gushing into the pool (Stotesbury et al. 2017). This caused the substance from the pool to be propelled upwards, creating the spatters of the wall and floor. These spatter patterns were documented in Fig. 4.3 photo.

Aside from the russet and red spatters on the wall and floor, the investigators found a russet and red pool containing a clot. Furthermore, numerous objects with spatter and low patterns were found. A curious object that is worth mentioning is a white basin covered with distinct low patterns. The shape and location of the patterns unequivocally indicate that they were created while the basin was in a different position from the one that was captured in the photo.

Fig. 4.1 The general layout of the crime scene, the restroom in this case

Fig. 4.2   The close-up of spatter patterns, low patterns, and their wipes, located on the left and back wall of the restroom

Fig. 4.3   The spatter patterns located on the floor, between the toilet and the entrance. The spatters were also found on the toilet itself, as well as on the basin laying nearby

Another interesting observation was the fact that even though the toilet is covered in spatter patterns, none were found behind the toilet. That fact indicates that the source of the spatter was located by the floor, between the entrance and the toilet. This once again confirms the version that the stains were created as a result of the substance dripping into the existing pool on the floor, causing spatters around it. In this case, the part of the spatters were blocked by the toilet itself. That's why the spatter patterns were found on the toilet, but not behind it. The state of the floor and the lower parts of the wall are documented in Fig. 4.4 photo.

An intense russet and red stain was found on rags and on the floor between the left wall and the toilet. On the wall above that pool, more spatter and low patterns were located. Most likely, the russet and red substance was gathering in this area, unable to spread to the sides, due to the presence of nearby objects. Furthermore, it appears that the majority of the substance had flowed down the wall onto the floor, which could be deduced due to the fact that in this area the stain is the most intense. The stains in this part of the toilet, both on the wall and on the floor, are documented in Fig. 4.5 photo.

The russet and red stains were also found inside of the toilet, as shown in Fig. 4.6 photo, indicating that their source must have been located over this surface as they were created. The toilet seat was up, and nothing was located on top of it. This is another piece of information that is crucial for determining the sequence of events.

There were some other interesting stains, aside from the spatter patterns. The contact stains located on the left side by the entrance are particularly important for reconstruction purposes. They were wipes, as indicated by their indistinct edges and

Fig. 4.4 The restroom with numerous russet and red stains, and a fragment of the common room, where the woman's body was moved

Fig. 4.5 The pool containing the clot, found between the left wall and the toilet, a fragment of the white basin with visible flow patterns, and spatter patterns with flow patterns on the left and front wall of the restroom

Fig. 4.6 The inside of the toilet with a visible russet and red stain, the white basin with visible flow patterns, and the fragment of the left wall with spatter and flow patterns

moderately regular shape. The stain's structure allowed the investigators to determine in which part it was more intense (Gardner 2002). It's worth mentioning that their location indicates that they were created when the door was open. The linear structure of the stain indicates the possibility that it had been created with a hand. It may also indicate the presence of fingerprints in the substance. This piece of information is crucial for determining the sequence of events (Fig. 4.7).

A stain that turned out to be the key to this case was the layout of spatters on the restroom door shown in Fig. 4.8 photo. The inner side of the door was covered in russet and red spatter patterns, up to 40 cm above the ground level. The spatters were round and oval-shaped and many of them had their tails pointing upwards, indicating that the area of origin was located below them, by the floor. Nearly identical spatters were found on the inner side of the doorframe. However, the most important piece of information was the lack of stains on the inner part of the doorframe, clearly seen as a white strip of the door's surface, with no spatters on it. Such a phenomenon is called a void. However, it is not just a lack of stains, because if it was, the majority of surfaces could be called voids. The most important aspect, and a determiner of whether a clean surface is a void or not, is its surroundings. In this case, the surface can be easily blocked or covered by itself, since as soon as the door is closed, the inner part of both the door and the frame is hidden. If the door is in this position, it is impossible for any substance to find its way into its inner part, which is only exposed when the door is open or at least ajar.

Fig. 4.7 Russet and red contact stains on the right side of the doorframe, as seen from the entrance

Fig. 4.8 Spatter patterns covering a large surface on the lower part of the door, with a clear exception of the void on the inner part of the frame and door

There was a single russet and red stain found on the doorframe, however, its nature is completely different from those around it. It was most likely created when the body was being moved through the open door.

The void on the inner side of the restroom door was easily noticeable when looked at from a 90 degrees angle, as seen in Fig. 4.9 photo. The void is a white strip of clean surface, found on the left side of the door, just by the hinges, stretching along the entire height of the spatter pattern. The investigators were not able to determine why was there a sanitary pad glued to the door, and deemed it a prison custom.

An important duty of the investigators, aside from documentation of bloodstain patters, is to note any information that may prove relevant to the case. In this crime scene, the presence of insects (flies) was noted (Choromański 2017). Although, in this case no insect stains that were impactful for the reconstruction of events were found, such things should be noted nonetheless. Insect stains are a result of an insect's contact with a substance and then spreading that substance onto other surfaces. The fly found in the restroom is shown in Fig. 4.10 photo. Study show that it is not easy to distinguish fly artifacts from regular human blood based on only presumptive tests (Durdle et al. 2015).

In the cell's common room, numerous contact stains were found on the floor by the victim. However, certain parts of her body were not covered in blood. Figure 4.11 photo shows the victim's right forearm with a noticeable cut on it, but no bloodstains surrounding it. This is a result of the medical examiner's activity, who had cleared that area, in order to categorize the wound for his report. The contact stains found by the victim's bloodied hands indicate that when the woman was being mover, the blood still was not clotted (Fig. 4.11).

Fig. 4.9 The lower part of the door, covered in numerous russet and red spatter patterns. The void, a white strip of clean surface, is visible on the inner part of the door

Fig. 4.10 Fly activity was documented on the inner side of the door, by the ventilation grid

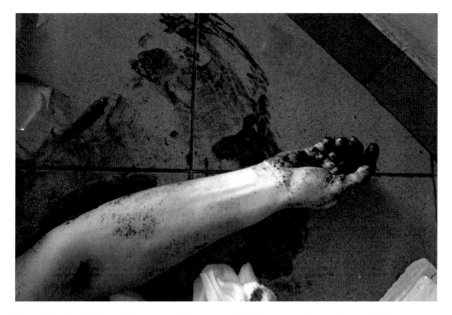

Fig. 4.11 The victim's right arm, and the russet and red contact stains on the floor in the common room, where the woman was temporarily moved to

Fig. 4.12 The dismantled head of the razor, along the razor blade used to inflict the wounds. Both items were found on the restroom floor

The tool used to inflict the wounds was a razor blade from the dismantled razor (Fig. 4.12). The investigators managed to find the fragment on the restroom floor.

4.4 Autopsy Findings

The woman's body was found with postmortem rigidity present all over the body. During the autopsy, a medical examiner found two significant wounds, the first was located on the right side of the neck, damaging the internal jugular vein and common carotid artery, a wound associated with a skin cut. The second was located on the right cubital fossa, damaging the median cubital vein, a wound associated with a double skin cut. Both were in the reach of the woman's hands, so they were described as potential self-inflicted wounds. Examiner found the damaged pad of a thumb in the woman's right hand. Medical expert concludes that the cause of death was exsanguination due to the neck and arm wounds. The toxicological analysis did not detect any drugs or alcohol in the woman's system.

4.5 Reconstruction of Event

A noninvasive examination of the three women remaining in the cell did not reveal any red and russet stains on them nor their clothes. Due to their isolation, while waiting for the police and the prosecutors to arrive, they had no opportunity to wash any stains off, implicating that the women had no contact with the victim or her blood. Further lab examination of their clothes confirmed that their clothes were clean. Smears taken from underneath their nails would also show no traces of blood.

The main place in which the event took place was the restroom. Numerous spatter patterns found on the wall indicate that a severe external hemorrhage was possible. There were also flow patterns present, indicating that the amount of blood on the surface was large enough to flow down the victim as a result of gravitational pull, despite blood's viscosity. It may indicate a severe hemorrhage if it was proven to be human blood.

The key piece of information in this case was the fact that nearly all bloodstains found on the crime scene, i.e.: the restroom, were created when the door was closed, as indicated by spatters on the frame and the door itself. Furthermore, no stains were found in the space between the framer and the door. If the door to the restroom was open, there would be spatters in that area.

The tool used to create the wounds was most likely the blade from the previously mentioned dismantled razor. Its small size made it difficult to locate, however, it was found on the restroom floor. It should be noted that the size of the wounds matched the size of the blade, making the object's use highly probable. Furthermore, the victim had wounds on her left-hand fingers, indicating that she could have held a small, unwieldy blade, with which she cut her fingers while inflicting other wounds upon herself.

Numerous spatters were found on the restroom's floor and the lower part of the wall, just above the floor. It is possible that they were created as a result of the subsequent drops dripping. It is also possible that the main source of the drips was the neck wound on the victim's neck, as well as her body, over which the blood might have been streaming down.

The shape and location of the spatters indicate that there were at least two locations areas of origin for the spatters patterns. Few spatter stains were chosen for detailed analysis (de Bruin 2011). It allows deducing that it is possible for a victim to be in two separate positions while creating the stains. In the first position, the victim was sitting on a toilet, leaning forward, and, as a result, the area of origin appears to be located closer to the door. In the second position, the victim is also sitting on the toilet, however, this time she is leaning back, with her back against the toilet and the wall, as indicated by spatters located in the corner of the room, between the wall and the toilet, where a large amount of the red and russet substance was found. Considering all the aforementioned information, and the fact that the majority of the red and russet substance was found in the corner, the following scenario may be assumed. It is possible that the victim was at first sitting on the toilet leaning forward, and wounded herself, which resulted in a hemorrhage. Once she transmitted the stains on surfaces from that position, it is possible she leaned back, possibly due to the loss of strength, while still sitting on the toilet. Then, she leaned against the wall and the toilet, while also tilting to her right. As a result of that, it is possible that the blood dripping from a wound accumulated in the corner of the room, in a form of an intense, russet and red clot, pool, and soaked clothing.

Furthermore, there were individual clear areas on the surface of the floor, created by objects shielding them from blood. Considering their location and size, it is highly probable that the victim's feet were in these spots.

Contact stains located on the walls of the restroom could have been created by the victim herself, or by the paramedics moving the body. There is no categorical evidence indicating who created them, thus, both hypotheses are equally believable.

As a result of the small space in which the victim was found, a scenario in which someone wounds her would result in an open door, and further stains in the main room. However, none such stains were found, and there were no signs of diluted stains, therefore, the hypothesis involving the possibility of cleaning or washing surfaces is absolutely unrealistic.

Characteristic contact stains, visible on the floor of the cell by the victim's hands, were created as a result of her transmission from the restroom to the main cell. The red and russet stains on her hands indicate that they had contact with the red and russet substance, meaning that it is possible that the wounds were self-inflicted. It's worth noting that there were red and russet spatters found on the victim's right hand, indicating that said hand was in the vicinity of the source of the spatter.

No evidence that could connect the events with a third party was found. There were no signs of dilution, drips moving away from the body, or any other signs characteristic for such cases. The only traces created by someone other than the victim, were most likely a result of the first aid, checking vitals, and moving her from the restroom to the cell.

4.6 Possible Areas for Mistakes

This case also presented opportunities for a bloodstain pattern analyst to make mistakes and overinterpret.

The first serious mistake that could have been made, would be a cursory examination of people residing in the cell. In such a case, every woman from the cell should be contained in a separate room, and there examined, with detailed pictures taken for documentation. Cursory examination and reporting would save time, however, such a shallow form of documentation would not be thorough enough to provide good working material for bloodstain pattern analyst.

Another mistake a bloodstain pattern analyst could make while attempting to recreate the crime scene and writing this case's report would be a misanalysis of the found spatter patterns. It should be noted that part of the spatter patterns, found both on the wall and the door of the restroom, indeed have a clearly defined shape, allowing to estimate the direction, from which droplets fell on the surface. However, it does not mean that all these spatter patterns should be taken under consideration when attempting to pinpoint the source of the spatter's location in space. A mistake in this case would be to consider spatter patterns, whose tails are pointing downwards. It is directly related to the laws of physics and the droplet's trajectory through the air. A spatter droplet usually moves along the path shaped similarly as the so-called cannonball trajectory; a parable that is slightly flattened on the side on which the droplet falls. Naturally, the droplet might move along a nearly straight line, however, this depends on spatial obstructions. A rectilinear motion can be observed as the droplet rises and falls, but obviously, these trajectories are but a segment of an estimated parable trajectory. Considering that, it is easy to determine that a scatter pattern whose tail is pointing downwards was created by a droplet that was falling. However, it could have been moving either in rectilinear motion, straight from the source of the spatter, or after having flown the trajectory of the estimated parable, in which case the source of impact is not that easy to determine. This means that a source of a spatter, whose tail is pointing downwards, can be above, on the same level or below the spatter pattern itself. Currently, research is being conducted on developing a model, allowing for a precise way to determine the source of the spatter, whose droplet moved along the parable trajectory. But at the time of writing both this case's report and this book, such methods are still not utilized in practice. The impossibility of determining the source of the spatter's location, based on the spatters pointing downwards means that they should not be examined for it, in order not to draw the wrong conclusions. Luckily, spatter patterns pointing upwards were also found on the crime scene. In such a case, the source of the spatter physically couldn't have been located above nor on the same height as them; it must have been located below. Based on them, you can attempt to locate the source of the spatter.

Another mistake you can encounter while reading an opinion prepared by mistrained bloodstain pattern analysts is overly interpreting spatter patterns in one, very specific, manner. It is a mistake to ascertain that a place in which a person was killed, is the same place in which a spatter pattern was found. A person's death should not be directly linked to a spatter pattern's discovery. What is interesting, the specialized literature does not mention this particular error, which I have encountered multiple times in practice. Perhaps the authors of professional publications, including bloodstain pattern analysts, fail to discuss it because they consider such interpretations ridiculous enough not to warrant mentioning it. Unfortunately, my experience unequivocally shows that this error occurs, and when it does, such information is of great use to misguided and unaware prosecutors and police officers. Why is this issue so important? In this particular case, it wasn't, however, you can imagine a scenario, in which a spatter was discovered in one place, and the body in another. In such a case, a question may be asked about the exact location, in which the person died. A mistrained bloodstain pattern analyst may, unfortunately, hypothesize that if a spatter was found, then it had to be the place in which the victim had died. It may be convenient for law enforcement authorities, however, it is not in any way supported by the scientific methodology. Drawing conclusions based on personal convictions is unethical and unprofessional. To sum up, the presence of a spatter, even an arterial one, does not imply that it was the place in which the murder occurred.

The next overinterpretation that can occur in a case like this one once again concerns spatter patterns. In the presented scenario, the medical examiner determined that the intactness of an artery in the victim's neck was compromised. For an analyst, such information is essential for detailed reconstruction. The examiner's finding allows a bloodstain pattern analyst to categorize certain stains as arterial spatters. If the medical examiner performed a less detailed examination of the body, in which the damage to the artery would be unmentioned, the analyst would be only able to categorize the stains more vaguely, as projected spatters. Luckily, in this scenario, the examiner performed a detailed corpse examination, allowing for a detailed reconstruction of the event. When writing an opinion, the rule is that it is better to make a vaguer statement, but one you're certain of than make a detailed and categorical statement, but with a hint of uncertainty.

Another thing worth noting is that none of the conclusions concerning the reconstruction are categorical. The reason for the fact that during this investigation no genetic examinations were performed. Thus, there is no confirmation that the substance found on the crime scene is human blood and you cannot categorically state to whose it is. Due to the lack of such information, the analyst may only draw conclusions based on a possibility, not a certainty. It indeed appears that in this place the only red and russet substance could be blood, however, it was not scientifically determined through biological and genetic tests. Such a statement is still just an unconfirmed hypothesis. That is why the bloodstain pattern analyst should not make categorical statements in a case like this one. It might seem irrational at first glance, but you should look at it from a broader perspective. After all, what is the difference between the presented scenario and the following? Let's assume that during the examination of clothes, a bloodstain pattern analyst found red and russet spatter patterns on the sleeve of the accused. If the analyst was to make a categorical statement that since the only one bleeding was the victim, the suspect must have covered himself with his or her blood, he would have committed an error of overinterpretation. After all, blood could have found its way onto his sleeve for numerous other reasons. As far as reconstruction of the event goes, both scenarios do not differ, and there is no reason to apply different reasoning to two methodologically similar scenarios. Consistency in the application of methodology allows for reproducible results.

Isolated drips were found on the crime scene. It is theoretically possible to calculate from what altitude did the drip fall, i.e.: how much distance did the free-falling drop of blood cover. However, in this case, there are two significant unknowns. The first one is determining certain pieces of data necessary for the calculations. The mathematical formula for determining the altitude from which the drop fell is as follows (Zadora 2014): $h = \frac{v^2}{2g}$, where $v = \frac{\eta}{\rho d^5}(2D)^4$, ρ is the density blood (1060 kg m^{-3}), η is its viscosity (4×10^{-3} Pa·s) and D is a diameter of the bloodstain. The only unknowns are h—altitude and d - diameter of the falling drop of blood. In practice, d is borderline impossible to estimate, since different objects and materials create drops in various sizes. Having two unknowns you cannot calculate the altitude from which the drop of blood fell with the use of this formula. Furthermore, the formula applies only for drips falling on the surface precisely at a 90 ° angle, and the surface on which it fell should be hydrophilic, flat, and non-absorbent. All these restrictions render this theoretically applicable formula unusable in practice. Unfortunately, in my career both as a scientist and a practitioner, I have encountered opinions, in which an expert was categorically stating the altitude, from which the drip occurred. In nearly all of those cases, such information was not even backed up by calculations, just the expert's personal experience. Such an approach is unacceptable. Sometimes experts rely on the presence of a so-called "crown" on the trace to determine the altitude. The crown is a wavy border of the stain, created from the increased kinetic energy of the drop, due to a fall from a larger altitude, resulting in a less regular, wavy boundary. Such a phenomenon does take place, however only when applied to one particular surface, a flat sheet of paper. The stain's shape drastically changes when the drop falls on a different surface, such as concrete, wood, plastic. On uneven surfaces, varying in absorptivity, the crown may be created even when the drop falls from a much lower altitude. Furthermore, not all drops are even. A 1 μl drop of blood will have different kinetic energy than a 10 μl, or a 50 μl one, and different kinetic energy will result in different characteristics of the stain it creates on any given surface. These facts are not usually considered by analysts using their personal experience, which is another mistake. To sum up, in specific scenarios it is theoretically possible to calculate the altitude from which the drop of blood fell, however, it's not applicable in practice. The analysts that precisely and categorically state in their opinions the altitude from which the drip occurred, make a serious mistake.

The previous paragraph also raises questions of how does such estimation contributes to the case. Its value for reconstructive purposes is not as big as one might think. Indeed, it might be relevant in attempts to assess the way by which the bleeding person left, the position he or she was in, or his or her estimated height. However, it is a field for further miscalculations and unverifiable hypotheses, such as determining the precise location from which the drip occurred. Therefore, it appears that the information concerning the altitude from which the drop of blood fell is not a contributing factor in the assessment of the legal classification of the act, thus it should not be a subject of all this attention. Perhaps further research in that area will allow for better reconstructions, ones that could be useful in a criminal trial.

4.7 Conclusion

Suicide in prison. It's not the first case were razor was used (Rautji et al. 2004; Akay 2017). Also it's not first time where suicide was perform in prison (Slade and Forrester 2015; Awenat et al. 2018; Voulgaris et al. 2019; Phillips 2020). After analysis, the matter may seem obvious. However, at the very beginning, there was no such certainty. Fortunately, in this case, an analyst of bloodstains along with technicians and investigators, was involved in the work on the scene from the very beginning. The essential evidence, in this case, was the void pattern revealed on the door and doorframe of the toilet door. This evidence clearly indicated that the door to the room was open at the time of creating the spatter. This led to a furthermore detailed inference. The toilet room area was too small for more than one person to move freely in it, so it was possible to exclude the participation of third parties at the time of the spatter on the door. Additionally, a detailed examination of the people presents confirmed the hypothesis that the roommates did not actively contribute to the woman's death. In fact, traces of possible blood outside the bathroom have been found. However, thanks to the knowledge and experience of the bloodstain pattern analyst working at the scene, it could be concluded that these traces could have appeared as a result of the actions of third parties on the spot, after the death of the inmate. These stains were of a contact nature, swipes and wipes. A highly possible way of their creation was the fact that the corpse was taken from the toilet to a shared cell in order to be examined by a forensic physician. This course was later confirmed by paramedics and the medic himself during the interviews.

The obvious is obvious because they are explained. A logical and clear explanation of the traces, in this case, contributed to a faster resolution of the case. In this case, no charges were brought against any person. The case was declared a suicide and then discontinued.

At this point, you can ask. What would happen if there was no analyst on the scene? What would happen if a blood trail analyst was not involved in any way? Would the course of events and the process itself have been the same? Unfortunately, it is impossible to answer these questions with a definite answer. One thing is certain. Thanks to the bloodstain pattern analysis, most of the question marks that stood in front of the investigators were dispelled in this case. The questions were answered to such an extent that the matter seemed almost trivial. However, as a practitioner, I would like to point out that there are no banal matters. Each case is different. Each is equally important. Each one should follow the same regime of work, inference, proving and documentation of traces and evidence. Each case, be it a murder or a suicide, should be treated with highest priority because in both cases, we are dealing with the loss of human life, which determines the seriousness of the proceedings. Analyzing bloodstains is not the golden mean. It is not a magic tool that solves all procedural and forensic problems. It is only an area of forensics that helps in conducting searches. The work regime of the analyst makes it necessary to keep specific documentation of traces in a strictly defined regime. An expert in this field draws attention to spaces that are often overlooked or ignored by investigators and experts from other areas. The bloodstain pattern analysis is not the alpha and the omega. It is only a help, which in some situations may turn out to be very important for the efficient work of investigators and law enforcement agencies.

References

Awenat YF, Moore C, Gooding PA, Ulph F, Mirza A, Pratt D (2018) Improving the quality of prison research: a qualitative study of ex-offender service user involvement in prison suicide prevention research. Health Expect 21(1):100–109. https://doi.org/10.1111/hex.12590

de Bruin KG, Stoel RD, Limborgh JCM (2011) Improving the point of origin determination in bloodstain pattern analysis. J Forensic Sci 56(6):1476–1482. https://doi.org/10.1111/j.1556-4029.2011.01841.x

Choromański K (2017) Bloodstain Pattern Analysis and enzymes *Enzymologia w obliczu wyzwań i możliwości XXI wieku*, (Wydawnictwo Naukowe TYGIEL sp. z o.o.: Lublin, 2017) ISBN 9788365598479

Durdle A, Mitchell RJ, Oorschot RAH (2015) The use of forensic tests to distinguish blowfly artifacts from human blood, semen, and saliva. J Forensic Sci 60(2):468

Gardner RM (2002) Directionality in Swipe Patterns. J Forensic Identification 52(5):579

Gardner RM, Maloney M, Rossi C (2012) A crime scene investigator's method for documenting impact patterns for subsequent off-scene area-of-origin analysis. J Forensic Identification 62(4):368

Phillips J (2020) What should happen after the death of a probationer? Learning from suicide investigations in prison. Probation J 67(1):65

Rautji R, Behera C, Kulshrestha P, Agnihotri A, Bhardwaj D, Dogra T (2004) An unusual suicide with a safety razor blade—a case report. Forensic Sci Int 142(1):33–35. https://doi.org/10.1016/j.forsciint.2003.12.020

Slade K, Forrester A (2015) Shifting the paradigm of prison suicide prevention through enhanced multi-agency integration and cultural change. J Forensic Psychiatry Psychol 26(6):737–758. https://doi.org/10.1080/14789949.2015.1062997

Stotesbury T, Taylor MC, Jermy MC (2017) Passive Drip Stain Formation Dynamics of Blood onto Hard Surfaces and Comparison with Simple Fluids for Blood Substitute Development and Assessment. J Forensic Sci 62(1):74

Voulgaris A, Hartwig S, Konrad N, Opitz-Welke A (2019) Influence of drugs on prison suicide—A retrospective case study. Int J Law Psychiatry 66. https://doi.org/10.1016/j.ijlp.2019.101460

Akay T (2017) Treatment of a patient who attempted suicide by swallowing a razor blade: a case report. Turkish J Colorectal Disease 27(3):97–99. https://doi.org/10.4274/tjcd.13008

Zadora G (2014) Application of bloodstain pattern analysis in the reconstruction of events. Problems of Forensic Sci 100:295–306

Chapter 5
Current Problems of Bloodstain Pattern Analysis

Abstract Development and technology have a huge impact on the shaping of forensics. In this chapter, I will try to present some of the most current problems for the community of bloodstain pattern analysts. Also, I will present compelling, current research to facilitate the work or solve these problems.

Keywords Bias · Interpretation · 3D scaning · Chemiluminescence

As technology advances, crime changes. Fortunately, investigators' tools are also developing. For a forensic science, which bloodstain pattern analysis is, there is a single principal theme: the blood itself behaves the same, regardless of the level of technology. The only effect that technical development can have is in the form of chemicals. Those substances can affect the physicochemical properties of blood. However, this issue is beyond the scope of this particular study. For over 100 years since the bloodstain pattern analysis was first used, the technology has developed beyond recognition. Previously, scientists could only observe stains; today, they can follow the formation of a spot thanks to high-speed cameras (Geoghegan et al. 2017; Williams et al. 2019). Every year we, as a scientific community, solve a new problem, a problem that in the past was impossible to overcome. A perfect example is the extensive use of photography to document bloodstain patterns nowadays. In the past, investigators would only have to rely on a sketch.

However, despite such a favourable situation, there are still things that cannot be overcome. Problems that today's scientists are trying to answer. I will try to present some of them in the following paragraphs.

5.1 Stating the Time of Creation of Stain

Over the last few years, research has been conducted on attempts to determine the time of the formation of the blood trail (Weber and Lednev 2020; Smith et al. 2020). This is of great importance, because with this type of information, you can locate an event on the timeline. Thanks to this, the reconstruction of the event is much more

K. Choromanski, *Bloodstain Pattern Analysis in Crime Scenarios*,
SpringerBriefs in Forensic and Medical Bioinformatics,
https://doi.org/10.1007/978-981-33-4428-0_5

detailed and can contribute a lot more to the case. Knowing that a given trace was created in a specific time period, it is much easier to exclude the participation of some potential suspects. As I mentioned before, there have been numerous attempts to reflect on this issue.

However, none has proven effective and is not used in practice to date. But scientists are not giving up. This is best evidenced by new research published in recent years, which has enormous potential for application in the practice of everyday investigative work. Some include the use of infrared spectroscopy (Kumar et al. 2020), others Raman spectroscopy (Doty et al. 2016; Menżyk et al. 2020). Therefore, it seems that in this area, over the next few years, the analysis of bloodstains will gain beneficial tools for conducting more detailed analyses.

5.2 Interpretation of Chemiluminescence Effect

In this book, I showed the effect of chemiluminescence when luminol is reacting with blood present on the surface of the evidence. But what if the forensic community is using luminol in a wrong way? What if some experts are over-interpreting the results of this chemical test. Those and other related questions will be answered by "PRELUDIUM 16" research project, ref. no UMO-2018/31/N/HS5/01031, financed by the National Science Centre, Poland.

In the pop culture, there is this premise, that luminol which will react with bloodstain giving chemiluminescence effect, specifically with washed stains. Forensic literature often calls this type of stains as hidden or secrete. It indicates purposeful activity done by a person to disguise these patterns. It can be misleading because small amounts of blood, that are not visible to the human eye, can also be created in different ways, like touching the surface with bloodied hands.

In science, there is a distinguishment between stains that are visible to the human eye, which is the main interest of bloodstain pattern analysis. There were numerous articles concerning the use of chemiluminescence, but there is no article regarding the interpretation of these enhanced stains found on a crime scene in any other way then washed.

The primary hypothesis for this project will be proving that: "There are differences between the way chemiluminescence looks on washed stains and the stains left passive". So far, current studies and practice as bloodstain pattern analysis expert show that this statement is true, but I want to prove it scientifically.

The second aim of this project is to establish what parts of today's crime scene investigation system can lead to some misleading conclusions, based on chemical enhancement of blood and how new methodology can overcome them. I think that the problem concerning overinterpretation of bloodstains is more profound then we believe. Numerous projects are focused on making chemiluminescence effect better, but there isn't any a paper that focuses on the validation of procedures of enhancing bloodstains. The project will focus on this part of the practice, and it will result in the creation of a new methodology of bloodstains, enhanced by chemiluminescence.

The research conducted in last year gives very interesting results. It turns out that understanding how luminol reaction really works is not obvious, even among the experts. Some of them claim that luminol reveals washed bloodstains, ignoring the fact that traces invisible to the human eye exist; stains that have never been washed off, and which will react with chemiluminescence effect. This already shows that this project may be of fundamental importance for the theory, but also for the practice of forensics. And it can have an international scope. The results of the project can give a new perspective of understanding criminal cases and bloodstain pattern analysis conclusions. The outcome will open new ways to review criminal cases in the context of evidence law concerning bloodstains enhanced by chemiluminescence, and may save the lives of innocents.

5.3   3D Documentation

As it has already been described, objective, detailed and precise documentation of the crime scene, as well as traces and evidence on it, are necessary to successfully resolve a criminal case. Until now, the documentation appeared mainly in the form of photographs, drafts and reports. But with every year we can see the intensive development of 3D scanners. They are becoming more significant, both in forensics (Raneri 2018; Wang 2019) and in the bloodstain pattern analysis (Hołowko et al. 2016). Currently, the 3D scanners, in addition to documenting the place of the incident and traces, have specialized software attached to them. Those programs allow, for example, to determine the area where a blow could be inflicted, resulting in the formation of a spatter (Hakim and Liscio 2015; Dubyk and Liscio 2016; Esaias et al. 2020). This type of solution greatly facilitates the work of bloodstain pattern analyst, and thus investigators. It seems to me that the coming years will bring more and more to 3D technology. This will improve crime scene investigation documentation process, make it easier and safer, specially during pandemic Coronavirus Covid-19 (Choromanski 2020) . It's only a matter of time when producents of hardware will make scanners cheaper and more accesible to a wider group of experts and police forces.

5.4   Psychology Research to Improve Pattern Analysis

This book indicates the essence of inference in the analysis of blood traces. I have also very often mentioned issues related to the overinterpretation and bias. This issue affects not only the analysis of bloodstains, but also other areas of forensics (Nakhaeizadeh et al. 2014; Sunde and Dror 2019; Thompson and Scurich 2019; Mattijssen et al. 2020a, b; Morling and Henneberg 2020; Hamnett and Dror 2020). Numerous studies related to this issue are currently underway. Some of them concern the issue of determining the direction of movement while creating the rubbing (Yuen et al. 2017). Others attempt to answer the basis on which the traces are actually categorized while tracing eye movement (Arthur et al. 2018). Additionally, it is investigated under what conditions the classification of traces may be difficult and why (Taylor et al. 2016). The impact of additional information on the expert's conclusions are also verified in terms of data management and the quality management process in units (Osborne and Taylor 2018). It seems that in the near future, issues of over-interpretation will continue to be the subject of research by scientists.

References

Arthur RM, Hoogenboom J, de Bruin KG, Green RD, Taylor MC (2018) An eye tracking study of bloodstain pattern analysts during pattern classification. Int J Legal Med 132(3):875

Choromanski K (2020) Performing bloodstain pattern analysis and other forensic activities on cases related to coronavirus diseases (COVID-19). Int J Legal Stud 1(7)::13–24

Doty K, McLaughlin G, Lednev I (2016) A Raman, "spectroscopic clock" for bloodstain age determination: the first week after deposition. Anal Bioanal Chem 408(15):3993–4001. https://doi.org/10.1007/s00216-016-9486-z

Dubyk M, Liscio E (2016) Using a 3D laser scanner to determine the area of origin of an impact pattern. J Forensic Identification 66(3):259–272

Esaias O, Noonan GW, Everist S, Roberts M, Thompson C, Krosch MN (2019) Improved area of origin estimation for bloodstain pattern analysis using 3D scanning. J Forensic Sci 65(3):722–728. https://doi.org/10.1111/1556-4029.14250

Geoghegan PH, Laffra AM, Hoogendorp NK, Taylor MC, Jermy MC (2017) Experimental measurement of breath exit velocity and expirated bloodstain patterns produced under different exhalation mechanisms. Int J Legal Med 131(5):1193

Hakim N, Liscio E (2015) Calculating point of origin of blood spatter using laser scanning technology. J Forensic Sci 60(2):409–417. https://doi-1org-10000b5r207ed.han.buw.uw.edu.pl/10.1111/1556-4029.12639

Hamnett HJ, Dror IE (2020) The effect of contextual information on decision-making in forensic toxicology. Forensic Sci International: Synergy. https://doi.org/10.1016/j.fsisyn.2020.06.003

Hołowko E, Januszkiewicz K, Bolewicki P, Sitnik R, Michoński J (2016) Application of multi-resolution 3D techniques in crime scene documentation with bloodstain pattern analysis. Forensic Sci Int 267:218–227. https://doi.org/10.1016/j.forsciint.2016.08.036

Kumar R, Sharma K, Sharma V (2020) Bloodstain age estimation through infrared spectroscopy and Chemometric models. Sci Justice. https://doi.org/10.1016/j.scijus.2020.07.004

Mattijssen EJAT, Witteman CLM, Berger CEH, Brand NW, Stoel RD (2020) Validity and reliability of forensic firearm examiners. Forensic Sci Int 307. https://doi.org/10.1016/j.forsciint.2019.110112

Mattijssen EJAT, Witteman CLM, Berger CEH, Stoel RD (2020b) Cognitive biases in the peer review of bullet and cartridge case comparison casework: a field study. Sci Justice 60(4):337–346. https://doi.org/10.1016/j.scijus.2020.01.005

Menżyk A, Damin A, Martyna A, Alladio E, Vincenti M, Martra G, Zadora G (2020) Toward a novel framework for bloodstains dating by Raman spectroscopy: How to avoid sample photodamage and subsampling errors. Talanta 209. https://doi.org/10.1016/j.talanta.2019.120565

Morling NR, Henneberg ML (2020) Contextual information and cognitive bias in the forensic investigation of fatal fires: do these incidents present an increased risk of flawed decision-making? Int J Law Crime Justice 62. https://doi.org/10.1016/j.ijlcj.2020.100406

Nakhaeizadeh S, Dror IE, Morgan RM (2014) Cognitive bias in forensic anthropology: Visual assessment of skeletal remains is susceptible to confirmation bias. Sci Justice 54(3):208–214. https://doi.org/10.1016/j.scijus.2013.11.003

Osborne NKP, Taylor MC (2018) Contextual information management: An example of independent-checking in the review of laboratory-based bloodstain pattern analysis. Sci Justice 58(3):226–231. https://doi.org/10.1016/j.scijus.2018.01.001

Raneri D (2018) Enhancing forensic investigation through the use of modern three-dimensional (3D) imaging technologies for crime scene reconstruction. Australian J Forensic Sci 50(6):697–707. https://doi-1org-10000b5r207ed.han.buw.uw.edu.pl/10.1080/00450618.2018.1424245

Smith FR, Nicloux C, Brutin D (2020) A new forensic tool to date human blood pools. Scientific Reports 10(1):1–12. https://doi.org/10.1038/s41598-020-65465-4

Sunde N, Dror IE (2019) Cognitive and human factors in digital forensics: problems, challenges, and the way forward. Digital Investigation 29:101–108. https://doi.org/10.1016/j.diin.2019.03.011

Taylor MC, Laber TL, Kish PE, Owens G, Osborne NKP (2016) The reliability of pattern classification in bloodstain pattern analysis, Part 1: Bloodstain patterns on rigid non-absorbent surfaces. J Forensic Sci 61(4):922

Thompson WC, Scurich N (2019) How cross-examination on subjectivity and bias affects jurors' evaluations of forensic science evidence. J Forensic Sci 64(5):1379

Wang J et al (2019) Virtual reality and integrated crime scene scanning for immersive and heterogeneous crime scene reconstruction. Forensic Sci Int 303, Oct. 2019, p. N.PAG. EBSCOhost, https://doi.org/10.1016/j.forsciint.2019.109943

Weber AR, Lednev IK (2020) Crime clock—analytical studies for approximating time since deposition of bloodstains. Forensic Chemistry 19. https://doi.org/10.1016/j.forc.2020.100248

Williams EMP, Graham ES, Jermy MC, Kieser DC, Taylor MC (2019) The dynamics of blood drop release from swinging objects in the creation of cast-off bloodstain patterns. J Forensic Sci 64(2):413

Yuen SKY, Taylor MC, Owens G, Elliot DA (2017) The reliability of swipe/wipe classification and directionality determination methods in bloodstain pattern analysis. J Forensic Sci 62(4):1037

Printed in the United States
By Bookmasters